MATHEMATICS

TEACH YOURSELF BOOKS

MAGIC SQUARE

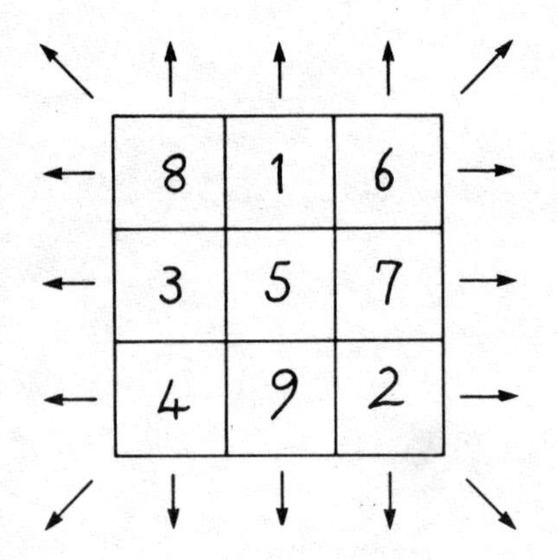

Worn by the people of London during the Great Plague, curious plaques of silver such as this were thought to possess supernatural powers. Because the numbers add up to 15 in every direction, the credulous believed that they would ward off the bubonic plague. Such magic squares had been familiar to mathematicians for countless generations.

MATHEMATICS

L. C. Pascoe, MA (Oxon.)

TEACH YOURSELF BOOKS

Long-renowned as the authoritative source for self-guided learning – with more than 30 million copies sold worldwide – the *Teach Yourself* series includes over 200 titles in the fields of languages, crafts, hobbies, sports, and other leisure activities.

British Library Cataloguing in Publication Data
Pascoe, L. C.
Mathematics – (Teach yourself books)
1. Mathematics – 1961 –
I. Title II. Series
510 OA37.2

Library of Congress Catalog Card Number: 92-80857

First published in UK 1983 by Hodder Headline Plc, 338 Euston Road, London NW1 3BH

First published in US 1992 by NTC Publishing Group, 4255 West Touhy Avenue, Lincolnwood (Chicago), Illinois 60646 – 1975 U.S.A.

Typeset by Macmillan India Ltd, Bangalore..
Printed in England by Cox & Wyman Ltd, Reading, Berkshire.

Impression number 22 21 20 19 18 17 16 15 14 13
Year 1999 1998 1997 1996 1995 1994

Contents

Introduction

In every previous work by the present author there were clearly defined pathways which could be followed, one leading to another, until a straightforward treatment of the subject was completed, in each case. The studies were identified by the titles. The present book is, however, altogether a different proposition. The reason is not difficult to find, for it is implied by the name itself, *Mathematics*. It is a vast subject which, like a full-grown tree, has many branches, although only a few of these are seen by the average citizen. How, then, should one proceed?

It took quite a time to come to a conclusion, but finally a plan of campaign crystallised. This was to begin with an historical background, and thence to progress through a fairly traditional basic course of study of the better known branches of mathematics, although sometimes deviating into interesting but lesser-known topics. It was highly desirable that this study should ultimately become unified and, at the same time, should be shown genuinely to possess practical value. There existed one suitable topic which could meet both of these requirements and, furthermore, could give the whole work contemporary viability. It was the study in depth of the operation and use of a reasonably equipped, but inexpensive, electronic calculator.

The book has very largely adhered to this scheme. The only difference is that, as experience is gained in the handling of the calculator by the reader, the opportunity has been given to him to undertake rather more searching, but really useful, studies. The machine, chosen about a year ago, was of American design, although made in Italy. It cost about £10 at the time and it is still in excellent working order, after very considerable use. The LED (Light Emitting Diode) read-out (or display) was chosen, rather than the slightly later LCD (Liquid Crystal Display) system. The former is very easy to read in poor light; the latter certainly is not, although it has a definite advantage in battery-saving.

Chapter 1 is a synopsis of mathematical thinking from its very beginnings, perhaps 10 000 years ago, to the seventeenth century AD. Of necessity, it is very much an outline, but it is hoped that it will prove interesting and instructive.

After this, there follow six chapters (2 to 7) which cover basic processes – and some a little more advanced – in arithmetic, algebra and geometry. They constitute approximately one-half of the book and they have been carefully selected to link up with the second half which, as indicated above, is intended to develop a real understanding of the uses of an electronic calculator.

The second half embraces Chapters 8 to 11. In the first of these, following a brief discourse on the background of the modern calculator, a typical keyboard is illustrated and explained, with various worked examples using different keys and combinations of them. There are numerous exercises for the reader. Chapter 9 deals with applications to the mensuration of curved surfaces and solids. Subsequently, Chapter 10 makes use of a calculator in connection with financial problems and loans, leading finally to a more advanced section on mortgages. Chapter 11 consists of a simple development of trigonometry via the use of similar triangles, and, apart from graphical considerations, almost all numerical work involves the calculator.

One word of advice with regard to mathematics. One *cannot* just read it and thereby hope to master the subject. As far as possible, everything should be personally worked out on paper, including the illustrated examples. The reader who is sufficiently self-confident should read these examples and try to answer them before looking at the solutions presented.

In conclusion, my grateful thanks go to Miss Janet Bolton, for the considerable help she gave with regard to the typing of the manuscript, and to Messrs Hodder and Stoughton, with whom I have now completed some twenty-five years of association.

L. C. Pascoe

1

A Brief History of Mathematics, from Earliest Times to 1660 AD

1 Preamble

I well remember, when I was young, that my mother adopted a somewhat eccentric method of reading a book. She invariably started at the last page or, on special occasions, the last two pages. This puzzled me greatly, and eventually it was impossible for me to refrain from the obvious question:

'Why are you reading the last page before the rest of the book?'

'Because I then know all about what is going to happen at the end, and I need not worry about it.'

This seemed an unexciting approach, although possibly beneficial to one's blood pressure. There is, however, one certainty: a study of mathematics cannot be successfully pursued like this. One must be prepared to proceed section by section, from the beginning, working as many of the demonstrated examples, as well as the exercises for the reader, as possible. In this way it is hoped that a sound groundwork of elementary mathematical *reasoning*, as well as ability to apply standard methods and formulae, will be acquired.

2 Mathematics in ancient times

We shall begin by putting the clock back by some 10 000 years, when the Upper Palaeolithic (late Old Stone) age, which had existed from about 18 000 BC, was merging into the Neolithic (New Stone) age. At around this time there appears to have existed a gentleman whom we

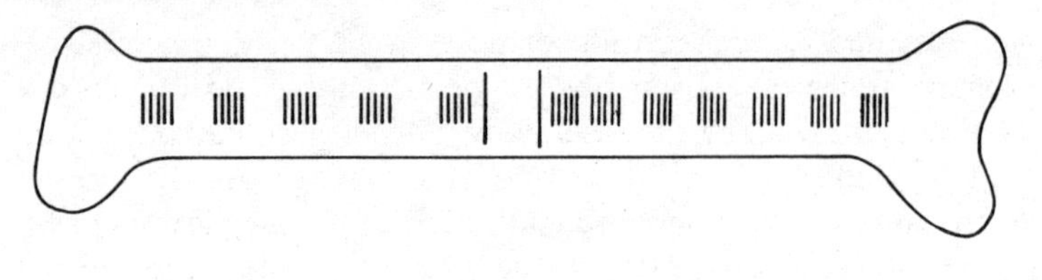

Fig. 1
The earliest known tally stick.
The marks were cut into the bone.

shall call Ug. He was truly remarkable as he made the earliest known tally stick, which was excavated in Moravia in 1937. It is sketched in Fig. 1, but there are no pretences as to anatomical accuracy!

The tally stick is a bone taken from a wolf, and clearly shows that, at this very early date, numbering in fives was not restricted only to the use of the human hand. One must naturally consider the possibility that Ug was outstandingly brilliant in his day, perhaps having a contemporary status equivalent to that held by Newton in the field of seventeenth-century (AD) mathematics. Even in the nineteenth century AD there existed, in some backward areas of the world, people who could only count

One, Two, One-Two, A Lot.

For all I know, there may still be such cases. Anyway, progress for the next few thousand years after Ug seems to have been limited.

We now move to the second millennium (2000 BC to 1000 BC) before the arrival of Christ, and we look towards the East. From a mathematical standpoint there were four countries to the fore – Babylonia, Egypt, India and China. Although these countries developed individually in many ways from late Neolithic times, their presentation of mathematics had numerous features of similar type. The countries did not, however, develop the subject at the same rate. Babylonia lay on a busy intersection of trade routes, and would have gained from wide exchanges of ideas; Egypt tended to be isolated; India will be referred to later, as early records were not kept on durable materials; China was insular, if for no other reason than sheer inaccessibility for most people – anyway China did not seem to wish to associate with other races.

Mathematics in Babylonia and Egypt, initially in strictly practical form, exhibited an appreciation of some of the basic processes of arithmetic and mensuration (the art of measuring and calculating length, area and volume, etc., of a body), together with an interest in astronomy (intertwined with astrology, no doubt). As the second millennium BC proceeded, so did mathematical thinking lead to basic algebraic ideas, and it also reached the threshold of primitive geometry. In 1858, there occurred the famous discovery of the Rhind Mathematical Papyrus, this being the work of the Scribe, Ahmes (*c.* 1650 BC). It gave a clear insight into the state of Egyptian mathematical knowledge at that period. It was not until the 1930s, however, when several archaeologists translated numerous Babylonian clay tablets, that the advanced state of Babylonian mathematics, as compared with Egyptian, came to light.

Less clearly are we able to assess the mathematical attainment in China, although in that vast country appreciable knowledge certainly existed contemporaneously with Babylonia and Egypt. Apart from the fact that, as in India, writing was upon perishable media, Imperial China was seized in 221 BC by Shih Hwang Ti, the Ch'in dynasty, emperor who ordered that all existing Chinese books be destroyed. Later attempts were made to rewrite earlier wisdom, but as always in such cases information was incomplete. Nevertheless, he unified China, gave it its name, built the Great Wall, and developed efficient government.

Writing materials used were as follows:

Babylonia: writing was by stylus used on wet clay tablets, which were then baked and were very durable indeed; the characters formed were cuneiform (wedge-shaped);

Egypt: writing was by pen on papyrus, which survived remarkably well in the dry climate;

China: writing was apparently on bamboo; furthermore, bamboo 'rods' were used to show numbers; they did not survive;

India: writing was mainly on palm leaf; this also did not survive from this period (although somewhat later examples exist).

Fig. 2 shows the states of numeracy in the second millennium BC, and we can compare the number systems of Babylon, Egypt and China.

1 The Babylonians possessed a symbol for 100, but they normally used a *base* of 60 (the sexagesimal scale) and *powers* of 60 (e.g. 60^2

Modern	1	2	5	6	9	10	12	20	50	60	100	1000	10 000
1 Babylonian (*c.* 1750 BC) (Cuneiform)	Y	YY	YYY YY	YYY YYY	YYY YYY YYY	‹	‹YY	‹‹	‹‹‹ ‹‹	Y	‹>		
2 Egyptian (*c.* 1650 BC) (Hieroglyphic)	|	||	||| ||	||| |||	||| ||| |||	∩	||∩	∩∩	∩∩∩ ∩∩	∩∩∩ ∩∩∩	𓍢	𓆼	𓂭
3 Chinese (*c.* 1500 BC) (Bamboo 'rods')	|	||	|||||	⊤	⊤||||	—	—||	=	≡||||	⊥	|	—	

Fig. 2 Ancient number systems (repetitive characters) compared with our modern ' Western ' (Hindu-Arabic) system.

$= 60 \times 60$). This method of counting corresponds to our own, in which we use the smaller base of 10. The Babylonian system had however certain disadvantages when compared with our own, but limitations of space will not permit detailed exposition here. Look, for example, at the number 9, requiring nine symbols! The symbol Y was used to represent 1 *or* 60 *or* 60^2, etc.

Consider a number 3852, and then the actual values of the 3, 8, 5, 2, respectively. *We* have a short way of writing $3 \times 10^3 + 8 \times 10^2 + 5 \times 10 + 2$ (where $10^3 = 10 \times 10 \times 10$ and $10^2 = 10 \times 10$).

In Babylonia we would have proceeded as follows, bearing in mind that $60^2 = 3600$:

$$3852 = 3600 + 252$$
$$= 1 \times 60^2 + 4 \times 60 + 12 \text{ (\textit{in this order})}$$

which leads to Y YY YY ‹YY

(notice the importance of the spacing).

2 The Egyptians possessed separate symbols for 1, 10, 100, 1000 and 10 000. They appear often to have written them from right to left, or sometimes in reverse order. It did not matter much – all the symbols for powers of ten were different! We shall, however, adopt the former (right to left) system. Let us look at

$$3852 = 3 \times 1000 + 8 \times 100 + 5 \times 10 + 2$$
$$= 2 + 5 \times 10 + 8 \times 100 + 3 \times 1000$$

In Egyptian, this would give quite a display:

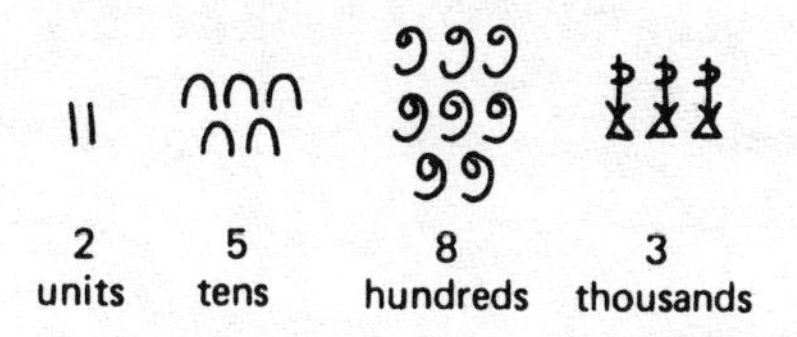

3 The Chinese system was easier in some ways, but it had the disadvantage that symbols could be repeated: the symbol for *one* (i.e. |) rotated on each multiplication by 10. (Refer back to Fig. 2.) For example:

1000 100 10 1 ; similarly 2000 200 20 2 ; *but* 6000 600 60 6, etc.

Thus $3852 = 3 \times 1000 + 8 \times 100 + 5 \times 10 + 2$ (in this order)

would have been written

It is straightforward to follow the procedure in these three countries, Babylonia, Egypt and China, provided that the relative positions of corresponding symbols are maintained. There was, however, no symbol for zero at that time, so a gap had to be left where necessary in Babylonian and Chinese numbers, but it was not necessary for those of Egypt.

Example 1 Rewrite the following in Hindu-Arabic (our modern) form:

(a) Babylonian (i) (ii) (iii)

(b) Egyptian (i) (ii) (iii)

We have

(a) (i) **13** (ii) $2 \times 60 + 10 =$ **130** (iii) $3 \times 60^2 + 43 \times 60 + 5 =$ **13 385**
(b) (i) $6 + 50 =$ **56** (ii) $1 + 10 + 200 =$ **211**
(iii) $300 + 2000 + 10\,000 =$ **12 300**

Example 2 Add the following ancient Chinese numbers, indicated by bamboo rods; translate these into modern numbers:

This example illustrates the importance of the position of the *second* number (||) in each case (a) and (b). In (a) it represents 200, and in (b) it represents 2.

(a) i.e.

$$\begin{array}{r} 468 \\ 200 \\ \hline \mathbf{668} \end{array}$$

(b) i.e.

$$\begin{array}{r} 468 \\ 2 \\ \hline \mathbf{470} \end{array}$$

EXERCISE 1

This is a useful exercise for developing mathematical thinking.

1 Write the following Hindu-Arabic numbers in the form used in Babylonia, but do *not* use the symbol for 100:

(a) 23 (b) 70 (c) 241 (d) 163 (e) 4082.

2 Put the following ancient Egyptian numbers into modern form:

(a) (b) (c)

3 Transcribe the following ancient Chinese numbers into modern form:

(a) (b) (c) (d)

4 Transcribe from Babylonian to ancient Egyptian:

(a) (b) (c)

3 The Greeks and the Romans

By the time that the first millennium BC was well under way, the balance of power in Eastern Europe and the Middle East had undergone a radical change. There had been many years of unrest, fighting and migration among the different nations. Cyrus II of Persia conquered Babylonia in 538 BC and Cambysis of Persia conquered Egypt nine years later, but the actual finale for the Pharaohs was in 525 BC, when Psammenitus was murdered. The mathematical leadership of Babylonia and Egypt was over.

To the fore there came the Hebrews, Phoenicians and Greeks, and it is from the last-named that we find the next important steps in mathematics. The Persians, tiresome fellows that they were, attempted to conquer the Greeks, who put up very stiff opposition. The Persians were defeated at Marathon (490 BC), Salamis, a great naval battle (480 BC), and finally at Plataea (479 BC). The Greeks were safe, but as a matter of fact their mathematics was well established appreciably earlier than this. Their victories, however, ensured that much of their knowledge survived for posterity.

The first notable Greek mathematician was Thales (*c.* 640 BC–*c.*546 BC). Originally a successful merchant, he laid the foundations of geometry based on logical reasoning, as compared with the earlier methods of trial and error or practical experiment. The Egyptians and Babylonians had merely used fixed sets of rules without any explanation. Now, however, the fundamental nature of *proof* and justification had arrived. Theoretical mathematics was alive.

Pythagoras (*c.* 581 BC–*c.* 497BC), who *proved* the famous theorem that 'In a right angled triangle, the square on the hypotenuse (the side opposite the right angle) is equal to the sum of the squares on the other two sides', was probably a pupil of Thales. Pythagoras, among other things, also founded a great school of philosophy which studied arithmetic, astronomy, geometry and music.

In the triangle ABC (Fig. 3), angle C (written $\hat{C}$) is a right angle. Then Pythagoras' theorem states that

$$AB^2 = AC^2 + BC^2$$

which can be written

$$c^2 = a^2 + b^2$$

where a, b, c are the lengths of the sides BC, CA, AB, respectively, and a^2 means $a \times a$.

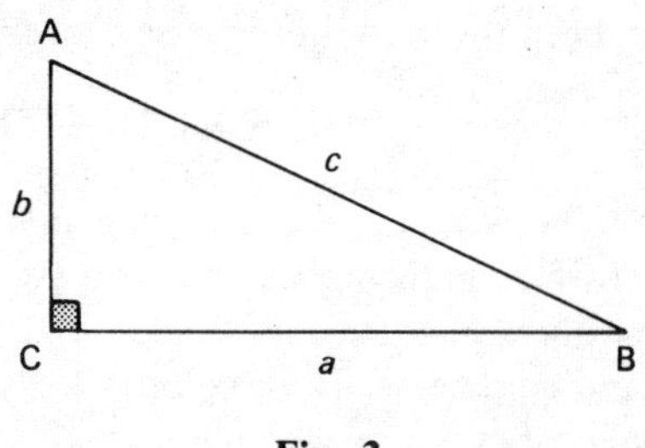

Fig. 3

The most important figure in Greek mathematics, and pre-eminent in the history of Pure Geometry, is surely Euclid, who flourished *c.* 300 BC. His great work, called *Elements*, was a concise record of the subject of his day. Although it contained other branches of mathematics, it is renowned for its geometrical content. Some of the theorems were Euclid's own, but many were obtained from other sources. He laid them all out in reasoned order. Countless thousands of children, many of whom did not bless his name, of necessity learned to wrestle with the intricacies of the proofs – even up to a mere thirty years ago! By this time, interest in pure geometry was beginning to wane as a result of greater emphasis on analytical methods, and because of the developing enthusiasm for modern mathematics and computing. One now tends to see only a few simple applications of pure geometry in a modern elementary course of mathematics.

Grecian mathematics received a severe jolt about 450 years before the birth of Christ. A philosopher, Zeno of Elea (*c.* 490 BC–*c.* 420 BC), formulated eight famous paradoxes – the most famous being *Achilles and the Tortoise* – which remained unanswered for about 2000 years. The basic question involved was whether or not the *sum* of an infinite number of quantities, however small, *must* necessarily be itself infinite or whether it *could* be finite. The Greeks never solved this, and their mathematical progress was consequently handicapped. Their systems of counting (there were three of them) did nothing to help, although the Ionic, which came last (probably *c.* 100 BC), was an improvement on the earlier forms. Here is an example from the Ionic, which used all 24 Greek letters plus three special ones, making 27 numerals in all:

$$\Delta\Pi\Omega = 800 + 80 + 4 = 884$$

It was written this way, but the *value* would not have been altered had it been reversed, in the case of a number less than 1000, for:

$$\Omega\Pi\Delta = 4 + 80 + 800 = 884, \text{ as before}$$

The Romans did not take much interest in the furtherance of abstract mathematical thought. They restricted themselves to practical mathematics, such as would be needed in civil engineering, building and commerce. Their country's system must be basically familiar to almost everyone who has seen an old clock, or the date of printing in an old book. The way in which the Romans wrote their numbers differed, however, in some ways from the adopted forms used much later in the Western world, particularly with regard to 9, 90, 900 and 1000.

For example:

Number (Modern Hindu-Arabic)	Number (Modern use of Roman)	Number (First century AD Roman)
9	IX	VIIII
1000	M	CIↃ
1980	MCMLXXX	CIↃDCCCCLXXX

The curious symbol CIↃ arose as an artistic development from O, a circle, representing 1000. This later became ↀ which split up as ᗡD, i.e. 500 written backwards and then forwards (500 + 500), or equally well as CIↃ! Anyway, it was later replaced by M (for *mille* = 1000 (Latin)).

4 The development of mathematics in the Western world

Let us first consider the development of our 'Western' numerals (1, 2, 3, 4 . . . 9 and, of course, 0); they are not really Western at all! Their name is Hindu-Arabic, and they seem first to have seen the light of day in India *c.* 250 BC, give or take the odd 50 to 100 years. By about 760 AD they had reached Baghdad. Modified by the Arabian peoples, they had by the sixteenth century become something like this:

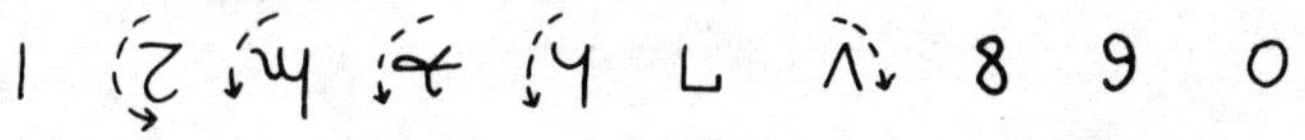

Fig. 4a

By rotating the 2nd, 3rd, 4th, 5th and 7th numeral as shown, we gain considerable resemblance to our current system (with the exception of number 4, which needs a little more imagination), as we can see below:

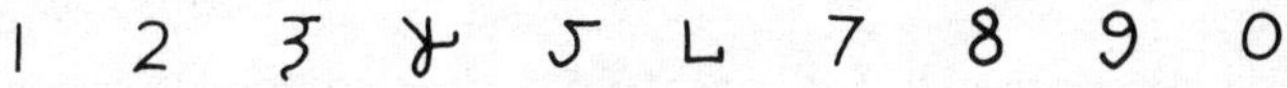

Fig. 4b

Incidentally, zero (0) may have appeared on the scene by approximately 500 AD, long after the other numerals were in use.

It has already been mentioned that basic ideas of algebra were known in Babylonia and Egypt. Solutions of algebraic equations of limited type were also undertaken at this early period. Knowledge of the subject was extended by Diophantus of Alexandria (fl. *c.* 250 AD) but the man who provided the name of the subject was Mohammed ibn Musa Al-Khowarizmi, in his treatise *Al-jabr w'al muqabalah* (*c.* 825 AD). Translated into Latin, it came to the West, where it was highly regarded; even so, there was still no *formal* presentation of algebra as we now know it. How then did the signs and symbols develop?

Let us firstly consider the signs for addition, subtraction, multiplication and division, which are the four basic processes of arithmetic – although nowadays they are also frequently to be found in algebra! The first two (addition and subtraction) were represented by and , respectively, as early as *c.* 1650 BC, in the Rhind Mathematical Papyrus, to which reference has already been made; these signs were later discontinued.

Our modern signs:

+ (plus: Latin *plus* = more) ADDITION SIGN

– (minus: Latin (neuter of *minor*) = less) SUBTRACTION SIGN

were first *printed* in 1489, by Widman, in Leipzig, although they were used earlier on sacks of grain to indicate that they were heavier or lighter than a standard weight. The third sign (for multiplication) is attributed to the English mathematician, William Oughtred (1575–1660), and seems to have been in use by 1631:

× (multiplied by: Low Latin *multiplus*) MULTIPLICATION SIGN

The fourth sign (÷) had a chequered history of meanings, but ultimately Rahn, a Swiss mathematician, adopted it to represent division; this was then popularised by English usage:

÷ (divided by: Latin *divisum* = separated, divided) DIVISION SIGN

In 1542, Robert Recorde (1510–58), another English mathematician, wrote *The Grounde of Artes*, a practical arithmetic, in *English*;

needless to say, it proved immensely popular, as it was no longer necessary for ordinary people in Britain to master Latin before studying mathematics. Recorde followed this work with (*inter alia*) *The Pathewaie to Knowledge*, a textbook of geometry, and *Whetstone of Witte*, which included algebra. He brought into use the sign of equality:

= (equals: Latin *aequalis*) SIGN OF EQUALITY

The development of decimals, which may have first appeared in India, *c.* AD 595, was extremely sketchy for many centuries. The earliest *printing* of the decimal point was apparently in the Pellos Arithmetic of 1492, but decimals were not systematically developed until Simon Stevin (or Stevinus) (1548–1620) took them in hand in Bruges, *c.* 1585.

It must be mentioned that the true name of decimals is *decimal fractions*. This is because they are obtained by converting fractions with any denominators into fractions whose denominators are powers of 10, namely 10, 100, 1000 and so on. For example:

$$\frac{1}{2} = \frac{5}{10} = 0.5; \quad \frac{1}{4} = \frac{25}{100} = 0.25;$$

$$\frac{2}{3} = \frac{666\ldots}{1000\ldots} = 0.666\ldots \simeq 0.667$$

(correct to three decimal places)

where the symbol $\simeq$ means 'approximately equals'. The decimal notation is merely a different way of writing a fraction with a denominator which is a power of 10, but the value of the decimal point lies in the ease with which one can manipulate numbers written in this form.

The actual use of *letters* as symbols representing numbers was originated by François Vieta (1540–1603), a French mathematician. The system appeared in 1591, in his important algebraic work *In artem analyticam isagoge* (An Introduction to Analytical Theory).

If the reader now checks back he will see that by 1659, after a tremendous struggle to evolve a handy system of numbers and basic symbols for arithmetic and algebra – a process that took about 3000 years! – mankind at last possessed a valuable mathematical shorthand, as follows:

1 The figures (digits) 1, 2, 3 . . . 9, together with 0, and their extension to numbers containing two, three, or more digits (e.g. 2507 is a four-figure number, i.e. one containing four digits).
2 Four basic signs for fundamental processes: $+$, $-$, $\times$, $\div$
3 A sign of equality: $=$
4 Algebraic letters (e.g. a could mean a certain number, and b could mean another number).
5 An alternative representation, in decimals, for numbers containing fractions, as and when required.

These lead to very much shortened relationships. Consider the examples:

(A) (i) When the number 7 is added to the number 9, the result is 16;
(ii) When the number 5 is subtracted from the number 9, the result is 4;
(iii) When the number 4 is multiplied by the number 8, the result is 32;
(iv) When the number 8 is divided by the number 4, the result is 2.

These longhand statements reduce to

$$\left.\begin{array}{ll}\text{(i) } 9+7=16 & \text{(ii) } 9-5=4 \\ \text{(iii) } 4\times 8=32 & \text{(iv) } 8\div 4=2\end{array}\right\} \quad \text{ARITHMETICAL RELATIONSHIPS}$$

(B) A calf costs £x and a lamb cost £y. Farmer Giles buys 4 calves and 7 lambs, at the stated price for each type of animal. He has to pay a total of £160. Express this as a single equation.

We note that 4 calves cost £$4x$ and 7 lambs cost £$7y$ and that the total cost is £160. Hence, removing the £ sign as it occurs throughout, we have

$$4x+7y=160 \qquad \text{AN ALGEBRAIC EQUATION}$$

The use of algebraic symbols in the above equation need not be fully comprehended at this stage. It is explained in Chapter 3, where the above-named gentleman is again active. After mastering some of the later work in this book, the reader is recommended to return to Chapter 1. He will then understand more clearly the logical development of basic mathematics, and how relevant this chapter is to a proper appreciation of the subject.

Historical notes on mathematical progress from 1660 onwards (which, coincidentally, marks the year in which the Monarchy was restored in Britain, and Charles II ascended the throne) will appear at various points in later chapters, whenever they may prove to be constructive.

5 A moral tale

Some years ago, when the present author was laying out the first edition of *The Encyclopaedia of Dates and Events* (Teach Yourself Books), a slightly embarrassing error occurred. Having adopted the use of a minus sign to represent any BC date (e.g. −55 means 55 BC, it was not necessary to use a sign for dates AD (e.g. 871 indicates 871 AD). One had, however, temporarily forgotten that dates on the calendar break a mathematical rule of continuity in numbering, with regard to the birth of Jesus Christ, and as a result this important date was entered as 0 AD. The main part of the encyclopaedia runs chronologically from −5000 to 1970. If we ascend a ladder of time, from shortly before the birth of Christ to shortly after (Figs 5a and 5b), we see an anomaly.

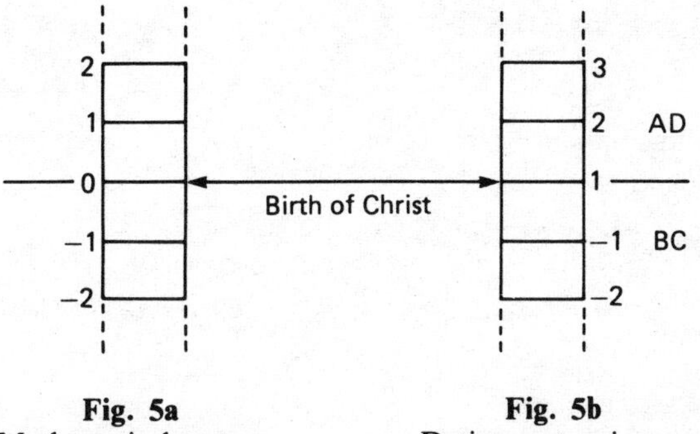

Fig. 5a
Mathematical system

Fig. 5b
Dating system in use

Mathematically, as we go up from −2 we proceed through −1, 0, 1 and 2, *but* there is no year 0 on the calendar! The birth of Christ immediately introduces 1 AD (the new era in the Christian world); hence a year has been lost (e.g. from 50 BC to 50 AD is 50 + 50 − 1 years, i.e. 99 years). This was because, as already mentioned, there was no symbol for 0 until about 500 years later.

It need hardly be added that, when noticed, the error was corrected, but it does not really improve our calendar, for it is nowadays generally accepted that Christ was probably born in 4 BC!

The story is given because it brings out very clearly the illogical results which can occur if one ignores the rules of mathematics, such as has arisen by taking 1 AD instead of 0.

2

The Basic Processes of Arithmetic

1 Introduction

Before we launch ourselves into mathematics in general, it is desirable to check briefly our recollections of the fundamental processes of arithmetic. This chapter is devoted to that end, for it will be found that many arithmetical rules apply to other branches of mathematics, or at least point the way to relevant methods. For the reader who feels, perhaps, that he is in need of a greater background of arithmetic, it may be helpful to mention that this can be found in *Arithmetic* (Teach Yourself Books), by the present author; the subject is there developed from its very beginnings.

2 The four rules

Arithmetic is the study of numbers under various operations, of which the basic ones are the 'four rules', namely, addition, subtraction, multiplication and division; among other relevant topics are fractions, factors, decimals, powers and the extraction of roots. The subject also includes applications to everyday life.

When stating a process to be carried out, it is always best, where possible, to be concise. Mathematics is an extremely concentrated shorthand, aimed at reducing long and sometimes complicated ideas to a manageable size. Look at the following:

'From the sum of 7 and 8 is subtracted 5'

which should be thought of as

'7 plus 8 minus 5'

and written as

$$7+8-5.$$

Addition

Example 1 (Adding in a row)

$15+7+38=\mathbf{60}$ $(5+7+8=20$; put down 0, carry 2)

$(1+0+3+2=6)$

Example 2 (Adding in a column: larger numbers) Add together 457, 8605 and 29.

We have

$$\begin{array}{r} 457 \quad (+) \\ 8605 \\ 29 \\ \hline \mathbf{9091} \\ 1\ 2\ \ \end{array}$$

Remember to put units under units, tens under tens, and so on.

Subtraction

As before, we can subtract in a row for easy numbers, and in a column for larger ones.

Example 3 Subtract 127 from 455.

We have $455-127=\mathbf{328}$ $(15-7=8;\ 5-2-1=2;\ 4-1=3)$

Example 4 What is the value of $2796-1908$?

We have

$$\begin{array}{r} 2796 \quad (-) \\ 1908 \\ \hline \mathbf{888} \\ 1\ 1\ \ \end{array}$$

Multiplication

4×7 *means* '4 lots of 7', i.e. $7+7+7+7=28$,
or '7 lots of 4', i.e. $4+4+4+4+4+4+4=28$ (the same result).

We multiply whichever way round is more convenient, for example:

$$9\times 6=6\times 9=54.$$

This can be stated as '9 times 6' or '9 multiplied by 6' or 'the *product* of 9 and 6'.

Example 5 (Short multiplication) What is the cost of 9 washing machines at £237 each?

We have

```
 237  (×)
   9
----
2133        Total cost £2133
 3 6
```

Total cost £2133

Example 6 (Long multiplication) Multiply 725 by 396, i.e. find 725 × 396.

We have

```
    725  (×)
    396
-------
  4 350
 65 25
217 5
-------
287 100
```

Division

Division is the process of sharing; it is the inverse of multiplication.

Example 7 (Short division) Divide 3276 by 12; i.e. 3276 ÷ 12.

We have

```
    8 3
12)3276
   ----
    273
```

Example 8 (Long division) Find the value of 4983 divided by 37.

```
    134
   ----
37)4983
   37
   ---
   128
   111
   ---
    173
    148
    ---
     25 ←——    remainder
```

The value is 134, remainder 25; as, however, the divisor was 37, the value can be written **$134\frac{25}{37}$**. Alternatively, in decimal form it is **134·68**, approximately.

Definitions

A *multiplicand* is a number to be multiplied by another number, called the *multiplier* (e.g. in Example 5 above, 237 is the multiplicand and 9 is the multiplier).

A *dividend* in this context is a number to be divided by another number, called the *divisor*; the *quotient* is the quantity which results from the division, and the *remainder* (if any) is what is left over (e.g. in Example 8 above, 4983 is the *dividend*, 37 is the *divisor*, 134 is the *quotient* and 25 is the *remainder*). The relationship between these terms is:

$$\text{Dividend} = \text{Divisor} \times \text{Quotient} + \text{Remainder}$$

In our example: $4983 = 37 \times 134 + 25$

3 Brackets

Numbers enclosed by brackets are to be considered as a whole. If more than one kind of bracket is in use at the same time, we start with the innermost. Although there are four standard types of bracket in general use, we shall generally use two, namely:

( ) *brackets* (more accurately called *parentheses*), and { } *braces.*

Brackets are of great importance in algebra and will be dealt with more fully when we reach that stage.

Example 9 Simplify (a) $8 \times (6+3)$, (b) $38-(16+15)$.

We have (a) $8 \times (6+3) = 8 \times 9 = \mathbf{72}$;

(b) $38-(16+15) = 38-31 = \mathbf{7}$

A useful, if unattractive-looking, mnemonic (memory aid) is the coined word BODMAS, which gives rise to a batting-order of operations, from 1 to 6 (wherein nowadays no. 2 has restricted use).

1	**B**	Brackets	Start with the insides of brackets (if any)
2	**O**	Of	(This used to be a powerful form of multiplication, preceding division)
3	**D**	Division	Carry out division next (if any)
4	**M**	Multiplication	Then multiplication (if any)
5	**A**	Addition	Then addition (if any)
6	**S**	Subtraction	*Finally* subtraction (if any)

Example 10 Simplify $6 \times \{7 + 45 \div (12 - 3)\} - 4(8 + 3)$.

We have

$$\begin{aligned} & 6 \times \{7 + 45 \div 9\} - 4 \times 11 \\ &= 6 \times \{7 + 5\} - 44 = 6 \times 12 - 44 \\ &= 72 - 44 = \mathbf{28} \end{aligned}$$

EXERCISE 1

1 (a) $17 + 21 + 6$, (b) $14 + 9 + 36 + 17$, (c) $128 + 37 + 79$

2 Add (a) $\begin{array}{r} 217 \\ 28 \\ \underline{466} \end{array}$ (b) $\begin{array}{r} 416 \\ 2937 \\ \underline{828} \end{array}$ (c) $\begin{array}{r} 12\,753 \\ 3\,609 \\ \underline{46\,778} \end{array}$

3 (a) Subtract 2196 from 3072,
(b) Find the value of $61\,843 - 57\,966$.

4 Find the value of (a) $217 + 593 - 678$, (b) $4073 - 2932 - 786$.

5 Multiply (a) 398 by 7, (b) 6745 by 29.

6 Divide (a) 273 by 3, (b) 2688 by 7, (c) 93 324 by 12.

7 Work out the following, and in each case give the answer in three forms, (i) with remainder, (ii) with a fraction after the whole number part, (iii) in decimal form, correct to 2 places of decimals:

(a) $816 \div 59$, (b) $2961 \div 147$, (c) $38\,479 \div 482$.

8 What is the value of $1234567 \times 9 + 8$? (Remember BODMAS!)

This is an interesting result and it is useful, because it gives a test for checking that an eight-digit display electronic calculator is in working order. For those readers who enjoy this kind of thing, there exist other expressions which can be made up (such as 12345679×8) in such a way as to yield amusing, but useful, answers.

9 Simplify (a) $32 - (18 + 7)$, (b) $17 + 3 \times (12 - 5)$,
(c) $3 \times (6 - 4) - 2 \times (9 - 6)$
(d) $3 \times (16 + 4 \times 7) - 5 \times (17 - 54 \div 6)$
(e) $208 - 2 \times \{27 - 56 \div (36 - 29)\}$.

10 A plane has to fly a distance of 2850 miles in six hours. After $3\frac{1}{2}$ hours' travel, it is found that the average speed made good has been 436 m.p.h. What must be the speed made good for the rest of the journey in order to arrive on time? (The speed made good is the speed over the surface of the earth; this is the one that is

important. It is derived from air speed and wind velocity, but that need not bother the reader!)

11 A young gentleman rejoicing in the name of Wilfred Bunter attends a boarding school. He receives a splendid circular cake from home and he invites seven of his friends to share it with him. Wilfred carefully cuts seven slices of cake, each having an angle of 30° at the centre, stating that 'this is to be fair to all'. Bearing in mind that a complete rotation is 360°, how many times bigger than each of his friends' slices is the remainder, which is Wilfred's piece?

After cutting three pieces each having an angle of 30°, Wilfred's greed overcomes him. Unnoticed by the assembled company, he reduces his other four friends' pieces to an angle of 27° each. How much of the whole cake does Wilfred get? (Give the answer as a decimal.)

4 Indices: powers of numbers

Where we have repeated multiplication involving the same number, we can use a shorthand notation. Consider $3 \times 3 \times 3 \times 3$, which can be written 3^4, where 4 is the *index* (plural, *indices*) or *power* (of the number, 3, in this case). Incidentally, as $3 \times 3 \times 3 \times 3 = 81$, then $81 = 3^4$. Other examples are $2 \times 2 \times 2 \times 2 \times 2 = 2^5$, and $5 \times 5 = 5^2$. Furthermore we can, if necessary, think of 5 as 5^1.

Addition and subtraction of indices

Consider $5^2 \times 5^6 = (5 \times 5) \times (5 \times 5 \times 5 \times 5 \times 5 \times 5)$, where the number 5 appears 8 times. We deduce that $5^2 \times 5^6 = 5^8$. This is interesting because $2 + 6 = 8$. Such a relationship is generally true, i.e. when *multiplying* together different powers of the *same* number, we *add* the indices. For example:

$$7^2 \times 7 \times 7^3 = 7^{2+1+3} = 7^6$$

Let us now look at $6^5 \div 6^3$, which is shorthand for

$$\frac{6 \times 6 \times 6 \times 6 \times 6}{6 \times 6 \times 6} = 6 \times 6 = 6^2.$$

Hence we have $6^5 \div 6^3 = 6^2$ and we notice that $5 - 3 = 2$; thus it would seem that, when *dividing* one power of a number by another power of the *same* number, we *subtract* the second power from the first. This also is generally true, for example:

$$2^3 \times 2^4 \div 2^6 = 2^{3+4-6} = 2^1 = 2.$$

Multiplication of indices

What happens if we have, say, $(3^4)^2$? We have already seen that any number multiplied by itself can be written as $(\text{number})^2$, e.g. $5 \times 5 = 5^2$, and so $5^2 = 5 \times 5$, hence

$$(3^4)^2 = 3^4 \times 3^4 = 3^{4+4} = 3^8.$$

Now $4 \times 2 = 8$, and this suggests that in such a case we multiply the indices. Taking another example:

$$(8^2)^3 = (8 \times 8) \times (8 \times 8) \times (8 \times 8) = 8 \times 8 \times 8 \times 8 \times 8 \times 8 = 8^6,$$

where $2 \times 3 = 6$. The result is generally true for whole number indices, for example:

$$(17^5)^4 = 17^{5 \times 4} = 17^{20}.$$

Division of indices is left until a later stage, as the ideas are rather more advanced (see Chapter 5).

5 Factors and prime numbers

Consider $6 = 2 \times 3$; the number 6 is said to have *factors* 2 and 3, which are whole numbers that divide *exactly* into 6. The only other factors of 6 are 1 and 6 itself. Hence all the factors of 6 are 1, 2, 3, 6. If we now look at 12 we find $12 = 1 \times 12 = 2 \times 6 = 3 \times 4$, and all the factors of 12 are 1, 2, 3, 4, 6, 12.

The numbers 6 and 12 each have several factors, but there are many numbers which have no factors other than unity (i.e. 1) and themselves, e.g. $5 = 1 \times 5$, $7 = 1 \times 7$, $11 = 1 \times 11$, $13 = 1 \times 13$; none of these has any other factor. Such numbers are called *prime*. The prime numbers less than 50 are as follows:

1, 2, 3, 5, 7, 11, 13, 17, 19, 23, 29, 31, 37, 41, 43, 47.

The number 1 is quite frequently omitted from the list of prime numbers, but at this stage we shall retain it.

We quite often need to factorise numbers, and this is best done by obtaining the *prime* factors. Taking an easy example,

$$24 = 2 \times 12 = 2 \times 2 \times 6 = 2 \times 2 \times 2 \times 3 \quad \text{(we cannot go further)}$$
$$= 2^3 \times 3 \quad \text{(using Section 4 above).}$$

Some numbers can be nasty, e.g. $3953 = 59 \times 67$, both of which are prime. We shall avoid these harder cases in the present book, except perhaps where a hard factor is actually given.

Example 11 Factorise (a) 1530, (b) 48 216.

(a)
```
2)1530
3) 765
3) 255
5)  85
    17
```

(b)
```
2)48 216
2)24 108
2)12 054
3) 6 027
7) 2 009
7)   287
      41
```

(a) We have $1530 = 2 \times 3 \times 3 \times 5 \times 17 = 2 \times 3^2 \times 5 \times 17$,
(b) We have $48\,216 = 2 \times 2 \times 2 \times 3 \times 7 \times 7 \times 41 = 2^3 \times 3 \times 7^2 \times 41$.

It will be observed that we divide by the suitable prime numbers in ascending order, starting with the smallest which will divide exactly into the given number.

6 A few simple tests for prime factors

1 Any *even* number, namely one ending in 0, 2, 4, 6 or 8, is divisible exactly by 2. Examples are 6, 18, 22, 74, 210.

Any number not divisible by 2 is called *odd*. Such numbers end in 1, 3, 5, 7 or 9. Examples are 9, 15, 127, 471, 543.

Numbers are alternately odd and even in counting, the even ones (as mentioned) all being divisible by 2.

ODD	1		3		5		7		9		11		and so on
EVEN		2		4		6		8		10		12	

2 Any number is divisible by 3, if the sum of its digits (figures) is divisible by 3. Examples:

(a) In 348, $3+4+8 = 15$; $15 \div 3 = 5$ exactly, therefore 348 is divisible by 3.
(b) In 329, $3+2+9 = 14$, *not* exactly divisible by 3, therefore 329 is not divisible by 3 exactly.

3 Any number is divisible by 5, if the given number ends in 5 or 0, but not otherwise. Examples: 35 and 270 are divisible by 5, but 71, 84 and 213 are not.

Tests for other factors such as 4, 6, 8 and 9 are given in *Arithmetic* (pp. 19–20). The harder tests for the prime factors 7, 11 and 13 and for all other factors up to 16 are to be found in *New Mathematics* (pp. 81–5); both are Teach Yourself Books.

7 Highest common factor (HCF)

The highest common factor of a group of numbers is the largest number which will divide into all of them. To find it in a particular case, we express each number of the group in prime factors.

Example 12 Find the HCF of 90, 255 and 345.

We have

$$\begin{aligned} 90 &= 2 \times 3 \times 3 \times 5 \\ & \qquad\quad \downarrow \quad \downarrow \\ 255 &= \qquad\; 3 \times 5 \times 17 \\ & \qquad\quad \downarrow \quad \downarrow \\ 345 &= \qquad\; 3 \times 5 \qquad\quad \times 23 \end{aligned}$$

These three numbers have one factor 3 and one factor 5 in common. Therefore HCF is $3 \times 5 =$ **15**.

8 Lowest common multiple (LCM)

The LCM of a number is the smallest number into which they will all divide exactly.

Although it may sound eccentric, the Lowest Common Multiple of a group of numbers is always larger than their Highest Common Factor (unless all the numbers are the same, which is a trivial case). The following example should make the reasoning clear.

Example 13 Find (a) the HCF and (b) the LCM of 56, 60 and 63.

(a) HCF of

$$56 = 2^3 \times 7 \qquad = 2^3 \qquad\qquad \times 7$$
$$60 = 2^2 \times 3 \times 5 \qquad = 2^2 \times 3 \times 5$$
$$63 = 3^2 \times 7 \qquad = \qquad 3^2 \qquad \times 7$$

There is *no factor common to all these numbers*, other than number 1, which is of little interest, i.e. HCF = **1**

(b) LCM of

$$56 = 2^3 \qquad\qquad \times 7$$
$$60 = 2^2 \times 3 \times 5$$
$$63 = \qquad 3^2 \qquad \times 7$$

LCM is $2^3 \times 3^2 \times 5 \times 7 = \mathbf{2520}$

Note that 2^2 and 2^3 will both divide into 2^3; also 3 and 3^2 will both divide into 3^2.

There are, of course, larger common multiples of 56, 60 and 63 (e.g. $2 \times 2520 = 5040$), but 2520 is the *lowest* one. This, alone, is of use among the common multiples.

EXERCISE 2

1 Factorise the following numbers, where possible, giving the answers as multiples of prime numbers in which indices (i) are not used, (ii) are used, where appropriate:
(a) 45, (b) 225, (c) 72, (d) 222, (e) 1323, (f) 3570, (g) 936, (h) 35 154, (i) 253) (this does factorise), (j) 241, (k) 289.

2 Find the HCF of each of the following groups of numbers:
(a) 36 and 48; (b) 70, 84 and 112; (c) 462, 825 and 1056; (d) 105, 180, 525 and 441.

3 Determine the value of the LCM of each of the following number sets:
(a) 9, 15; (b) 6, 8, 9; (c) 42, 60, 175; (d) 8, 10, 12, 14, 18.

4 A special pack of cards is to be made so that it can be used for any number of players up to six, each player receiving the same number of cards. What is the least number of cards needed?

5 Three services of buses leave a bus stop together, on time, at 9 a.m. The first service runs every 10 minutes, the second every 12 minutes and the third every 18 minutes. When next will they all be due to leave the bus stop together?

6 Two cars are being driven at constant speeds round a circular track. The first car completes one lap in 90 seconds and the second car does this in 105 seconds. At a particular moment the cars are side by side. How long is it before they are again side by side, and how many laps will the slower car have completed in this time?

9 Fractions

In this chapter we have so far only considered *integers* (whole numbers). We now investigate some of the properties of *fractions*, i.e. parts of whole numbers. Everyone is likely to be familiar with simple cases, such as $\frac{1}{2}, \frac{1}{4}, \frac{2}{3}, \frac{3}{4}$, and so on. These may of course be expressed in decimal form, namely 0.5, 0.25, 0.67 (approx.), 0.75, respectively, but we shall quite frequently find, when we come to the introduction to algebra which follows in the next chapter, that fractions are of greater value.

The fraction $\frac{3}{4}$ has an upper part (the *numerator*) indicating the number of parts we are taking, 3 in this case; and a lower part (the *denominator*) indicating the total number of parts, 4.

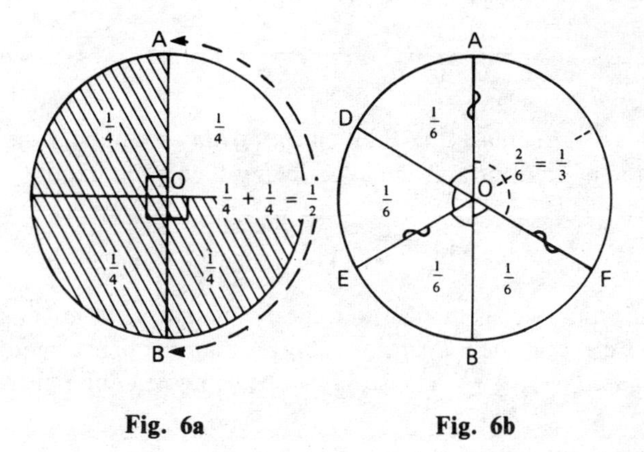

Fig. 6a **Fig. 6b**

In Fig. 6a, we see that the shaded part of the circle is $\frac{1}{4}+\frac{1}{4}+\frac{1}{4}$, i.e. $\frac{3}{4}$, whereas the unshaded part is $\frac{1}{4}$. The line AB passes through the centre, O, of the circle (i.e. AB is a *diameter*) and divides it into two equal parts. Hence, also, $\frac{1}{4}+\frac{1}{4}=\frac{1}{2}$.

In Fig. 6b, we see that the area OADEBF is $\frac{1}{6}+\frac{1}{6}+\frac{1}{6}+\frac{1}{6}=\frac{4}{6}$

whereas the other part OFA is $\frac{2}{6}$, but area OAE = area OEF = area OFA = $\frac{1}{3}$, and therefore $\frac{2}{6} = \frac{1}{3}$. Figs 6a and 6b are simple illustrations of the general rule which follows.

We may cancel out common factors in the top and bottom of a fraction. For example:

$$\frac{30}{48} = \frac{\overset{\overset{5}{\cancel{15}}}{\cancel{30}}}{\underset{\underset{8}{\cancel{24}}}{\cancel{48}}} = \frac{5}{8}$$

where we divide by the common factors 2 and 3 in turn. After some practice, those not already familiar with the method will divide more quickly by the common factor 6 (= 2 × 3), in this example.

Finally, reverting to Figs 6(a) and (b), we see that:

(a) $$\frac{3}{4}+\frac{1}{4}=\frac{3+1}{4}=\frac{4}{4}=1$$

(b) $$\frac{1}{6}+\frac{1}{6}+\frac{1}{6}+\frac{1}{6}+\frac{2}{6}=\frac{1+1+1+1+2}{6}=\frac{6}{6}=1$$

Definition: A fraction is the quotient (ratio) of two numbers, the upper being the numerator and the lower being the denominator:

$$\text{Fraction} = \frac{\text{Numerator}}{\text{Denominator}}.$$

If the numerator is smaller than the denominator, we have a *proper fraction*, e.g. $\frac{5}{6}$; if the numerator is bigger than the denominator, it is called *improper*, e.g. $\frac{11}{6}$. The latter can be expressed differently, when divided, thus:

$$\frac{11}{6}=\frac{6+5}{6}=\frac{6}{6}+\frac{5}{6}=1+\frac{5}{6}=1\frac{5}{6}$$

Similarly $\frac{20}{6} = 3\frac{2}{6} = 3\frac{1}{3}$, for $20 \div 6 = 3 +$ remainder 2. It is not, however, necessary to put down all these steps, and the reader should practise doing them mentally.

Example 14 What fraction of a day is 5 hr 20 min?

We have

$$\frac{5\,\text{hr}\ 20\,\text{min}}{1\ \text{day}} = \frac{5\frac{1}{3}\,\text{hr}}{24\,\text{hr}} = \frac{\frac{16}{3}}{24} = \frac{\frac{16}{3}\times 3}{24\times 3} = \frac{16}{24\times 3} = \mathbf{\frac{2}{9}}.$$

Firstly we convert to hours (20 min = one-third of 60 min); the hour, top and bottom, can be thought of as 'cancelled'; we now make the numerator into an improper fraction; we then multiply top *and* bottom by 3, which does not affect the ratio; $\frac{16}{3}\times 3 = 16$ (the numerator); we then cancel down using the common factor 8.
Alternatively,

$$\frac{5\text{hr}\ 20\,\text{min}}{1\ \text{day}} = \frac{(5\times 60+20)\text{min}}{24\times 60\,\text{min}}$$

$$= \frac{320}{1440} = \mathbf{\frac{2}{9}}, \text{ after cancelling down.}$$

EXERCISE 3

1 Find the number of centimetres in $\frac{3}{20}$ m.
2 How long is two-thirds of one day?
3 What fraction is 2 weeks 4 days of 3 weeks 3 days?
4 What is the ratio of 6 m 45 cm to 7 m 5 cm, as a fraction in lowest terms?
5 Jones earns £5430 per annum, and Smith is paid £84 a week during a year of 365 days. Express Jones's income as a fraction of Smith's, in lowest terms. (*Hint*: How much does Smith earn each day?)

10 Addition and subtraction of fractions

When two or more fractions have the same denominator, it is easy to add or subtract them, for example:

$$\frac{7}{12}+\frac{3}{12}=\frac{10}{12}=\frac{5}{6}; \qquad \frac{5}{7}-\frac{3}{7}=\frac{2}{7}$$

If, however, as is usually the case, the denominators are different, we find the LCM of the denominators. Consider $\frac{3}{8}+\frac{5}{6}$. The LCM of 8 and 6 is 24; also $8\times 3 = 24$ and $6\times 4 = 24$.

Now

$$\frac{3}{8}=\frac{3\times 3}{8\times 3}=\frac{9}{24} \text{ and } \frac{5}{6}=\frac{5\times 4}{6\times 4}=\frac{20}{24}$$

therefore

$$\frac{3}{8}+\frac{5}{6}=\frac{9}{24}+\frac{20}{24}=\frac{29}{24}=\mathbf{1\frac{5}{24}} \text{ (= 1.208 approx., if needed)}$$

Similarly

$$\frac{7}{12}-\frac{8}{15}=\frac{7\times 5-8\times 4}{60}=\frac{35-32}{60}=\frac{3}{60}=\frac{1}{20}$$

(*Method*: LCM of 12 and 15, i.e. $2\times 2\times 3$ and 3×5, is $2\times 2\times 3\times 5 = 60$; $12\times 5 = 60$ and $15\times 4 = 60$.)

Example 15 Simplify $\frac{7}{12}+\frac{5}{8}-\frac{4}{9}$.

The LCM of $12 = 2^2\times 3$, $8 = 2^3$ and $9 = 3^2$ is $2^3\times 3^2 = 8\times 9 = 72$

therefore

$$\frac{7}{12}+\frac{5}{8}-\frac{4}{9}=\frac{7\times 6+5\times 9-4\times 8}{72}=\frac{42+45-32}{72}=\mathbf{\frac{55}{72}}$$

which does not reduce further.

When there are whole number parts present in the *addition* and *subtraction* (of fractions), we deal with these parts separately. This does *not*, however, apply in the same way when *multiplication* and *division* are involved (see the next section).

Example 16 Simplify $4\frac{1}{6}-5\frac{2}{3}+1\frac{3}{5}$.

Here we have

$$4-5+1+\frac{1}{6}-\frac{2}{3}+\frac{3}{5}=0+\frac{5-20+18}{30}$$

$$=\frac{23-20}{30}=\frac{3}{30}=\mathbf{\frac{1}{10}}.$$

Note: it is important (as given in BODMAS) that we *add* before *subtracting*.

11 Multiplication of fractions

When we have two or more fractions multiplied together we can cancel out any factors which appear in both the top and bottom, as though the expression is all one fraction, for example:

$$\frac{3}{14} \times \frac{21}{36} = \frac{\overset{1}{\cancel{3}}}{\underset{2}{\cancel{14}}} \times \frac{\overset{\overset{1}{\cancel{3}}}{\cancel{21}}}{\underset{\underset{4}{\cancel{12}}}{\cancel{36}}} = \frac{1}{8},$$

because 3 and 7 are common factors of the whole numerator and whole denominator.

Fig. 7 will make the idea easier to understand. Consider $\frac{2}{3} \times \frac{3}{4}$; lay out twelve squares as shown, 3×4, representing the denominators of the first and second fractions.

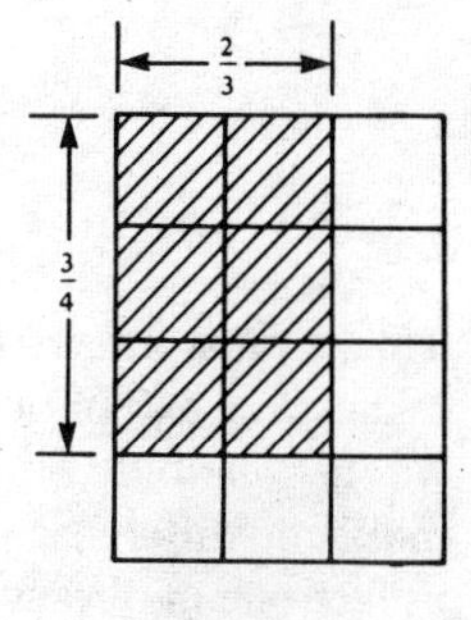

Fig. 7

Then

$$\frac{2}{3} \times \frac{3}{4} = \frac{2 \times 3}{3 \times 4} = \frac{6}{12} \left(\text{i.e. } \frac{6 \text{ squares}}{12 \text{ squares}} \right)$$

$$= \frac{1}{2}.$$

When some of the numbers to be multiplied consist of whole numbers combined with fractional parts, we convert these numbers to improper fractions and proceed as above.

Example 17 Simplify $\frac{3}{7} \times 2\frac{1}{2} \times 4\frac{1}{5}$.

We have

$$\frac{3}{\underset{1}{\cancel{7}}} \times \frac{\overset{1}{\cancel{5}}}{2} \times \frac{\overset{3}{\cancel{21}}}{\underset{1}{\cancel{5}}} = \frac{9}{2} = \mathbf{4\frac{1}{2}} \text{ (alternatively, 4.5)}$$

12 Division of fractions

Suppose we have

$$\frac{2}{5} \div \frac{8}{9} = \frac{\frac{2}{5}\text{(numerator)}}{\frac{8}{9}\text{(denominator)}}$$

As in section 10, above, if we multiply the top *and* bottom of a fraction by the same amount, we do not alter the value of the fraction. To simplify our expression, we multiply the top and bottom by the LCM of the *denominators* of $\frac{2}{5}$ and $\frac{8}{9}$, i.e. by 5×9 (as 5 and 9 have no common factor – other than 1). Therefore:

$$\frac{2}{5} \div \frac{8}{9} = \frac{\frac{2}{5} \times 5 \times 9}{\frac{8}{9} \times 5 \times 9} = \frac{2 \times 9}{8 \times 5} = \frac{2 \times 9}{5 \times 8} = \frac{2}{5} \times \frac{9}{8}$$

This is most interesting because it looks very similar to the original expression, except that the second fraction has been inverted and that we then *multiply* by it, instead of *dividing*. This is always true, for it makes no difference what the first fraction (or whole number) may be; in fact one may view the first fraction as a catalyst, needed to help a mathematical change to take place, in a way analogous to the action of a catalyst in chemistry (wherein the catalyst itself ultimately remains unchanged).

Anyway, to finish our example,

$$\frac{2}{5} \div \frac{8}{9} = \frac{2}{5} \times \frac{9}{8} = \mathbf{\frac{9}{20}}$$

In like manner, when dividing by a number such as $2\frac{1}{3}$, we convert it to an improper fraction, $\frac{7}{3}$, invert thus, $\frac{3}{7}$, and multiply by it.

Example 18 Simplify $3\frac{1}{3} \times 2\frac{4}{5} \div 1\frac{3}{4}$.

The expression is

$$\frac{10}{3}\times\frac{14}{5}\div\frac{7}{4}=\frac{\overset{2}{\cancel{10}}}{3}\times\frac{\overset{2}{\cancel{14}}}{\underset{1}{\cancel{5}}}\times\frac{4}{\underset{1}{\cancel{7}}}=\frac{16}{3}=5\frac{1}{3}.$$

Example 19 Simplify

$$\frac{\frac{3}{8}\times\frac{1}{4}+\frac{4}{5}}{\frac{5}{8}-\frac{1}{4}\times 1\frac{2}{5}}.$$

We first simplify the upper and lower parts independently.

Upper part

$$\frac{3}{8}\times\frac{1}{4}+\frac{4}{5}=\frac{3}{32}+\frac{4}{5}=\frac{3\times 5+4\times 32}{160}=\frac{15+128}{160}=\frac{143}{160}$$

Lower part

$$\frac{5}{8}-\frac{1}{4}\times\frac{7}{5}=\frac{5}{8}-\frac{7}{20}=\frac{5\times 5-7\times 2}{40}=\frac{25-14}{40}=\frac{11}{40}$$

∴ the expression is

$$\frac{\frac{143}{160}}{\frac{11}{40}}=\frac{\overset{13}{\cancel{143}}}{\underset{4}{\cancel{160}}}\times\frac{\overset{1}{\cancel{40}}}{\underset{1}{\cancel{11}}}=\frac{13}{4}=3\frac{1}{4}\text{ (or 3.25).}$$

Note: The original expression could have alternatively been written as $\left(\frac{3}{8}\times\frac{1}{4}+\frac{4}{5}\right)\div\left(\frac{5}{8}-\frac{1}{4}\times 1\frac{2}{5}\right)$.

EXERCISE 4

1 Find the value of $2518-924+146$.

2 From 1007 take the sum of 283 and 324.

3 From 2237 take the product of 65 and 34.

Simplify nos 4–9

4 $42-(9-2)\times 6.$

5 $28-102\div 12.$

6 $51-47+8\times 7.$

7 $30\div 4+1.$

8 $428-4\{3\times 17+4(10\div 2+3)\}.$

9 $\dfrac{7+12\div 2}{7\div 6+1}.$

Simplify expressions 10–17, giving the answers in lowest terms:

10 $\dfrac{2}{3}-\dfrac{3}{7}+\dfrac{1}{21}.$

11 $\dfrac{3}{5}-\dfrac{3}{7}\times 1\dfrac{1}{6}.$

12 $6\dfrac{1}{4}\div 3\dfrac{1}{3}-1\dfrac{1}{2}.$

13 $2\dfrac{1}{4}-1\dfrac{1}{5}-2\dfrac{1}{7}.$

14 $\left\{\dfrac{1}{3}-\dfrac{1}{10}\right\}\div\left\{\dfrac{2}{9}-\dfrac{1}{12}\right\}.$

15 $3\dfrac{3}{7}\times 4\dfrac{2}{3}.$

16 $\dfrac{\dfrac{3}{5}+\dfrac{2}{7}-\dfrac{1}{4}}{\dfrac{3}{5}+\dfrac{1}{7}-\dfrac{1}{4}}.$

17 $\dfrac{1\dfrac{3}{8}\times\dfrac{1}{4}}{\dfrac{5}{8}-\dfrac{1}{4}\times 1\dfrac{2}{5}}.$

18 Express £1.68 as a fraction of £2.73, in lowest terms.

19 A man dies and leaves $\frac{1}{6}$ of his estate to his son and $\frac{1}{7}$ of the remainder to his daughter. The residue is left to the widow and is valued at £24 000. What is the value of the son's legacy?

20 A cyclist is riding at $22\frac{1}{2}$ miles per hour. How fast is this, in feet per second? Assuming that 1 kilometre is $\frac{5}{8}$ mile, find the cyclist's speed in kilometres per hour.

21 Smith can erect a certain wall in 15 days, Jones can do the job in 12 days, and that highly skilled bricklayer, Robinson, in 10 days. How long should it take them to erect the wall if they work together? (*Hint.* Find firstly what fraction of the wall each can build in one day.)

The following questions do not necessarily require the use of fractions, but are intended to develop mathematical reasoning based on this chapter.

22 Robert, who does not have an electronic calculator with him, requires to multiply 298 476 by 79. When he has finished this he

notices, to his annoyance, that in error he has written down the first number as 298 467. Find by how much he has to correct his answer. (There is an easy way.)

23 Four bells are rung at intervals of 6, 8, 9 and 12 seconds, respectively. After starting together, how long is it before they again sound together?

24 How many numbers between 1 and 2500 contain *all* the numbers 2, 3, 4, 5, 6 and 7 as factors. (*Hint*: Find the smallest number into which 2, 3 . . . 7 will divide exactly. This number is called their . . . ?) Write down the required numbers.

25 A club had 280 members. Of these 196 play tennis, 244 are involved in the dramatic section and 160 play bridge. What is (a) the largest number, (b) the smallest number, of members who may be participating in all three activities? (Although this type of problem may be considered as associated with *set theory*, the example here is easily solved by arithmetic.)

26 Find the missing numbers in each of the following:

```
(a)     2 . 3 ×              (b)  5 . ) . . 2 . . ( . . 7
         . 8 .                           1 . .
       -------                          -------
        . . . 2                          1 . .
      . . . .                             . . .
      2 . .                              -------
      -------                             . . . .
      . . . 3 .  (Multiplication            . 1 .
                                          -------
                                              0  (Long division)
```

13 Tailpiece: hieroglyphic fractions

The Egyptians only used fractions having a numerator of 1, with the exception of $\frac{2}{3}$ (written as). In addition, no fraction of any one type could be repeated in the same expression (e.g. $\frac{1}{5}+\frac{1}{5}$ was not allowed). This made life difficult! The fraction $\frac{1}{2}$ had a special symbol (*gs*, meaning 'side'). All other fractions were represented as whole numbers under (*r*, meaning 'part'), thus:

$= \frac{1}{3}$; $= \frac{1}{12}$ (i.e. $\frac{1}{10+2}$)

Hence, the following number, taken from the Rhind papyrus and recorded in Sir Alan Gardiner's *Egyptian Grammar* (third edition, reprinted 1976) is, on reading from *left* to *right*,

$$= 5 + \frac{1}{2} + \frac{1}{7} + \frac{1}{14} = 5\frac{5}{7}$$

3

Algebra

1 What is algebra?

The usual definition of algebra is delightfully simple, namely that it is a generalisation of arithmetic. The statement is, however, rather limited, because algebra extends well beyond the normal bounds of arithmetic. It might be better to view the latter as an important subsection of the former. This may seem to be a quaint idea, as arithmetic came first, but, taking a real-life analogy, if we do not learn to walk properly before we run, we are likely to land in a horizontal position! Hence, basic arithmetical processes were examined in the last chapter, in preparation for extension. It should be mentioned, though, that algebra in its elementary form is no harder than arithmetic, once familiarity with the techniques is properly established.

Fundamental rules and processes of arithmetic frequently appear in algebra, sometimes in slightly modified guise, but the latter also requires additional signs, symbols and operations. Among those common to both studies are:

1 *The four rules*: addition, subtraction, multiplication and division; the signs + and − are used throughout algebraic work in this book, but × (for multiplication) will only appear infrequently, and ÷ (for division) will not often be needed; the symbol = (equals) is always used, when applicable.

2 *Rules for factors and fractions* (although the latter are, in general, more elaborate in algebra).

3 *Decimals*, for substitution in formulae and for solving algebraic equations which have terms with *numerical* coefficients.

4 *Ratio and proportion.*

5 *Rectangular co-ordinates for graphical work.*

6 *Brackets.* In Chapter 2, Section 3, it was stated that the contents of brackets must be treated as a whole. In arithmetic (as has been seen), we normally collect the contents together and *then* carry out any process required. In algebra, this is not always possible and we may have to apply the process to separate terms, for example:

$$\text{(i) } 3(2+7) = 3 \times 9 = 27$$
$$\text{(ii) } 3(2a+7b) = 6a+21b.$$

The fundamental difference between the two branches of mathematics is that whereas in arithmetic we use an ordinary number to indicate a *definite* value, in algebra we use a letter to represent an unknown or unspecified value. Compare the following:

(a) $6+6+6+6+6 = 5 \times 6$ (this has a known value)

with a corresponding generalised expression:

(b) $x+x+x+x+x = 5 \times x$ (this has an unknown value)

The first expression, (a), leads to the specific value, 30. The second expression, (b), can only lead to a specific result if we select a value for x; this may or may not be a whole number (e.g. x could be 257, 3.75, $\frac{4}{7}$, 0, -12, or any other number). If, say, we put $x = 3.6$, then $5 \times x = 5 \times 3.6 = 18$.

2 Omission of the multiplication sign in algebraic expressions

Consider the expression $5 \times x$ above, which is deliberately chosen because it illustrates the nuisance value of the multiplication sign ($\times$) when placed alongside the letter x, especially if typed or handwritten xx! The difficulty is avoided in algebra, as *we normally omit the multiplication sign between a number and a letter* (incidentally, the number is put first), for example:

$5 \times x$ is written $5x$

Likewise when multiplying two or more letters together, we also omit the multiplication sign, for example:

$$a \times b = ab; \quad p \times q \times r = pqr; \quad 3 \times a \times b = 3ab.$$

Slightly more subtle is the multiplication together of, say, $9p$, $\frac{2}{3}q$, $\frac{1}{4}r$ and s. Here we have

$$9p \times \frac{2}{3}q \times \frac{1}{4}r \times s = 9 \times \frac{2}{3} \times \frac{1}{4} \times p \times q \times r \times s = \frac{3}{2}pqrs,$$

where we simplify the numbers, when appropriate, as shown in Chapter 2. It follows that all of the multiplication signs are unnecessary in this case, although there are occasions when they are of value.

There is one difference: in arithmetic, an improper fraction such as $\frac{3}{2}$ would finally be given as $1\frac{1}{2}$ or 1.5. In algebra, however, we would leave, say, $\frac{3}{2}x$ as it stands, or express it as $1.5x$; we would *not* use $1\frac{1}{2}x$.

It is important to remember an invariable rule when applying algebraic letters to the solution of a problem: we must carefully define each letter. The following example illustrates a simple case.

Example 1 A piece of cloth is x metres long and y centimetres wide. What is its area, in square metres?

The length is x m and the breadth is y cm $= \frac{y}{100}$ m

$\therefore$ The area of the cloth, which is length $\times$ breadth, is

$$x \times \frac{y}{100} \text{ sq.m} = \mathbf{\frac{xy}{100}} \textbf{ sq.m.}$$

Note: The symbol $\therefore$, which means *therefore*, will be widely used.

3 Division

As with multiplication, division signs are only used sparingly in algebra. In such a case as $6x \div 15y$, it will be found helpful to put the expression in the form of a fraction:

$$6x \div 15y = \frac{6x}{15y} = \frac{2x}{5y}, \text{ after simplification.}$$

Where we have to divide by a fraction such as $\frac{a}{b}$, we proceed exactly as we did in arithmetic (p. 30). We invert the fraction and multiply by the result, for example:

$$x \div \frac{a}{b} = x \times \frac{b}{a} \boxed{= \frac{x}{1} \times \frac{b}{a}} = \frac{bx}{a}, \text{ which is } \frac{b}{a}x \text{ (cf. Section 2)}$$

The step enclosed by dotted lines is done mentally and is not normally written down.

Example 2 Simplify $\frac{3p}{4r} \times \frac{14qr}{pq} \div \frac{7q}{16}$.

Using the rule above, we have

$$\frac{3\not{p}}{\not{4}\not{r}} \times \frac{\not{14}\not{q}\not{r}}{\not{p}\not{q}} \times \frac{\not{16}}{\not{7}q} = \frac{3}{1} \times \frac{2}{1} \times \frac{4}{q} = \frac{24}{q}$$

(cancelling: $4r \to 1$, $14qr \to 2$, $pq \to 1$, $16 \to 4$)

After some practice the second step can be omitted.

4 Addition and subtraction

Consider the statement 'Farmer Giles has some sheep and some cows'. Can we represent this algebraically? Certainly, but it does not get us very far if this is all we know about Giles's animals. It is nevertheless instructive.

Let x be the number of sheep and y be the number of cows, then the total number of animals is $x + y$.

If, however, we *also* know that Giles has twice as many sheep as he has cows, we do not need *two* letters each representing a different number, for one letter is sufficient. In this case, let the number of cows be y, then the number of sheep is $2y$, and the total number of animals is $2y + y = 3y$. (In fact, we have replaced x by $2y$, i.e. we have used the equation $x = 2y$.)

Suppose now that Giles sells z of his sheep, then the number of sheep he has left is $2y - z$.

It will be observed that we cannot dispose of the + and − signs when they connect two or more *different* letters. This is an important variation from the procedure adopted for × and ÷ signs.

Example 3 Smith wishes to invest £36 000 in three types of security, namely building societies, British Funds and Ordinary Shares. His purchase of building society shares is three times as great as that of British Funds, which itself is to be twice as great as that of Ordinary Shares. How much money does he place in building societies?

Let the amount put into Ordinary Shares be £x, then the cost of British Funds is £$2x$, and the investment in building society shares is £$3 \times 2x$ = £$6x$. Now we know that all these acquisitions will cost Smith £36 000 (ignoring the brokerage and stamp duty on the purchase of British Funds and Ordinary Shares).

Hence

$$£x + £2x + £6x = £36\,000$$

Removing the £ signs and collecting the terms in x together (which we can do, as the same letter x is in each LHS term), we get

$$x + 2x + 6x = 36\,000$$

$$\therefore \quad 9x = 36\,000 \Rightarrow x = 36\,000 \div 9 = 4000$$

But the amount in building societies is £$6x$, i.e. £6×4000 = **£24 000.**

(The symbol ⇒ means 'implies that'.)

It is now time to summarise two rules for the use of algebraic letters:

1 Each letter used in a problem has a particular meaning, different from that of any other letter which may be used. Hence, we can say $3x + x = 4x$ *or* $5x - 2x = 3x$, for example, but we cannot simplify expressions like $3x + y$ *or* $5x - 2y + z - 6$ any further than their present forms, unless we have either additional information about relationships between the letters, or numerical values which may be substituted for them.

2 An algebraic letter represents a quantity, which *at that time* is not identified with any particular value. A specific value or values *may* however be given later and then substituted in each place where the letter appears in, say, a formula or result now found.

Example 4 below illustrates both these rules.

Example 4 Simplify the expression $x+4y-z-3y+5z-x+6$. Find also the value of the expression when $y=2$ and $z=1$.

We firstly sort out terms of the same kind, i.e. we separate those in x, y and z, and the independent number 6.

$$\begin{aligned}\therefore x+4y-z-3y+5z-x+6 &= x-x+4y-3y-z+5z+6\\ &= 0+y+4z+6\\ &= \mathbf{y+4z+6.}\end{aligned}$$

If $y=2$ and $z=1$ then,

$$y+4z+6 = 2+4\times 1+6 = \mathbf{12.}$$

We now see why the value of x was not quoted for the second part of the question. The terms in x have cancelled out, although this does *not* usually happen. Also, only after we have substituted for x and y can we collect the 6 with the other terms.

Example 5 Jones has £x from which he buys five dozen eggs at y pence a dozen. How much money has he left, in £s?

The cost of one dozen eggs is

$$y \text{ pence} = £\frac{y}{100}$$

$\therefore$ the cost of five dozen eggs is $£\dfrac{5y}{100} = £\dfrac{y}{20}$

$\therefore$ the amount of money left is

$$£x - £\frac{y}{20} = £\left(x - \mathbf{\frac{y}{20}}\right), \text{ which is neater.}$$

Example 6 The sum of the two numbers h and k is added to twice their product, the resultant expression being then divided by three times the first number. What is the result? (*Note*: the *product* is obtained by multiplying the numbers together.)

Twice the product of the numbers h and k is $2\times h\times k = 2hk$. Hence the total amount added up is $2hk+h+k$, but this is divided by $3h$,

$\therefore$ the result is $(2hk+h+k)\div 3h = \mathbf{\dfrac{2hk+h+k}{3h}}$

Example 7 A train is travelling at v m.p.h. How far will it go in 2 hr 12 min?

We have $\quad$ 2 hr 12 min $= 2\frac{1}{5}$ hr $= \frac{11}{5}$ hr

In 1 hr, the train travels v miles

In $\frac{11}{5}$ hr, it travels $\frac{11v}{5}$ **miles**

$\left(\text{alternatively written } \frac{11}{5}v \text{ miles}\right)$

Example 8 A rectangular room is of area A m^2 and it is of length l m. What is its breadth?

Let the breadth be b m, then as the area of a rectangle is length $\times$ breadth, we have A (m^2) $= lb$ (m^2) and hence $lb = A$, on changing sides; now dividing both sides by l, as in arithmetic, we have

$$\frac{\not{l}b}{\not{l}_1} = \frac{A}{l} \Rightarrow b = \frac{A}{l}, \text{ i.e. the breadth is } \frac{A}{l} \textbf{ metres.}$$

EXERCISE 1

1 Simplify the following expressions:

(a) $2 \times x \times z$ (b) $4 \times 7y$ (c) $3a \times \frac{1}{4}b$ (d) $\frac{3}{5} \times \frac{1}{4}a \times \frac{8}{9}b \times 6c$.

2 Express the following in simplest form:

(a) $2z \div 3$ (b) $15y \div 12$ (c) $21ab \div 14c$ (d) $2p \div pq$

(e) $15abc \div 35acd$ (f) $\frac{5}{14} \times \frac{3a}{8bc} \div \frac{10ab}{21c}$ (g) $\frac{3p}{8} \div \frac{9pq}{12} + \frac{1}{2q}$.

3 Simplify where possible:

(a) $5x + 6 - 4x$ (b) $5x - 7y + 6x + y$ (c) $3a - 6 + 2c - 9b$

(d) $5x - 7y + 12z + 9 - 3x - 7y - 4 + 4y$

(e) $8m + 7 - 9n + 4 - 5m + 10n - 12 - n - 3m$.

4 1 kg is approximately equal to 2.2 lb. How many ounces are there in n kilogrammes? Find, also, the number of grammes there are in x ounces.

5 The sum of three numbers is s. The first number is a and the second is b. What is the third? What is the *average* value of each

of the three numbers? (The average value of any one of a set of numbers is their total value divided by the total number of numbers present.)

6 A number r is multiplied by 3, and the result is divided by t. What is the answer?

7 The product of two numbers p and q is divided by their sum. What is the value obtained?

8 In a factory, x employees are paid £p each and y employees are paid £q each, for a week's work. Find the weekly wages bill.

What is the *average* weekly wage paid to an employee? (*Hint*: Divide the total wages bill by the total number of employees.)

9 A widow dies and leaves an estate valued at £x, after death duties and other expenses have been paid. Four cousins each receive £ y. The residue of the estate is to go entirely to the deceased's only son. Assuming that there is something left for him, what is the value of his share?

10 If $\dfrac{a}{2b} = c$, what is the value of b in terms of a and c? (*Hint*: Multiply both sides by $\dfrac{b}{c}$.)

11 Solve the equations (i) $4x = 7$, (ii) $3y + 2 = 23$, (iii) $z + \frac{1}{7}z = 16$

12 A train travelled m miles at u m.p.h. and then n miles at v m.p.h. How long did the whole journey take? What was the average speed for the journey?

5 Some definitions

We have reached the stage where we need some definitions:

1 A combination of letters and numbers, such as $4x$, y (which is $1y$) $\frac{1}{3}ab$, $27pqt$, i.e. one which includes multiplication and/or division, but does not include + or −, is said to be a single term or *monomial*. (This word really has a mixed ancestry! Greek *monos* = single; Latin *nomen* = name.)

2 Where we have a combination like, say, $3m + 5$, $2yz - x$, or $1 - \frac{2}{5}c$ we say that it consists of two terms, i.e. it is *binomial* (now *all* Latin! *bis* = twice, *nomen* = name).

3 For three terms (e.g. $a+b+c$, $y-2x-3z$, $3pq+2p-6$) we use the name *trinomial.*

4 For more than three terms, the name is *polynomial* (Greek *poly* = much/many).

5 Each combination of letters/numbers separated by + or − is called a *term,* the name having already been used in the above definitions.

6 All the algebraic groups illustrated above (three examples: $4x$, $a+b+c$, $3m+5$) are called *expressions.* An expression is a name given to any symbolic mathematical form, of which one type is the polynomial. We shall, in this book, only touch on straightforward cases.

6 An important algebraic law and an introduction to the use of brackets

It is very widely known that the area of a rectangle is the product of its length and its breadth, i.e. its area is length × breadth *or* breadth × length.

Let us look at the expression al. This may be regarded as the area, in square units, of a rectangle ABCD, of breadth a units and length l units (Fig. 8a). N.B. The rectangle is formally defined in Chapter 4 (definition 17a) and its deducible properties are given later in the same chapter.

In Fig. 8b, which represents the same rectangle ABCD, we put $l = x + y$ and on AB take the point E such that AE $= x$; it then follows that EB $= y$. Similarly, on DC we take the point F such that DF $= x$

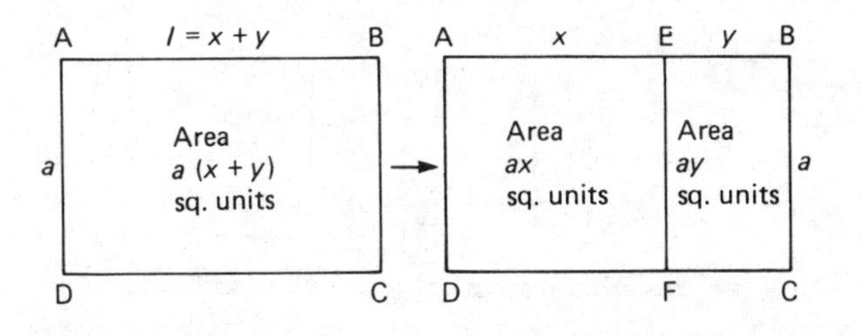

Fig. 8a Fig. 8b

(i.e. $FC = y$). We join EF, dividing the rectangle ABCD into two smaller ones AEFD and EBCF.

We shorten the phrase 'area of the rectangle' to the single abbreviation 'rect.' and we see that

$$\text{Rect. ABCD} = \text{AD} \times \text{AB} = a(x+y) \text{ sq. units} \qquad \text{Fig. 8a}$$

$$\left.\begin{aligned} \text{Rect. AEFD} &= \text{AD} \times \text{AE} = ax \text{ sq. units} \\ \text{Rect. EBCF} &= \text{BC} \times \text{EB} = ay \text{ sq. units} \end{aligned}\right\} \text{Fig. 8b}$$

But

$$\text{Rect. ABCD} = \text{Rect. AEFD} + \text{Rect. EBCF}$$

Hence $a(x+y)$ sq. units $= ax$ sq. units $+ ay$ sq. units, where the square units are self-cancelling along the line, so we get the important result:

$$\boldsymbol{a(x+y) = ax + ay.}$$

The above demonstration is *not* a rigid proof, but it is a good example of inter-relationship among different branches of mathematics, in this case between geometry and algebra. (See Law 5, below.)

Let us check the above equation in a simple arithmetical case, where $a = 7$, $x = 3$ and $y = 8$. We use the abbreviations LHS and RHS for 'left hand side' and 'right hand side', respectively, of an equation.

We have $\quad \text{LHS} = a(x+y) = 7(3+8) = 7 \times 11 = 77$

and $\quad \text{RHS} = ax + ay = 7 \times 3 + 7 \times 8 = 21 + 56 = 77$

$\therefore \quad \text{LHS} = \text{RHS}$

and the formula is checked in this instance.

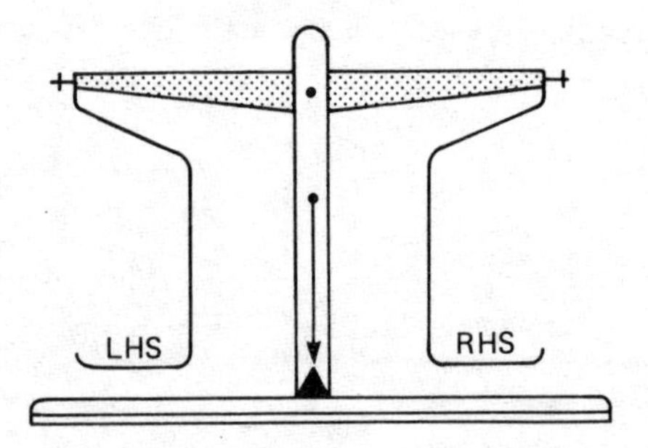

Fig. 9

The LHS/RHS method, of separately reducing two sides of an equation to a result which is seen to be common to both, is quite widely used in mathematical proofs or verifications.

In fact, every mathematical equation is analogous to a statement that we have a pair of scales on which the left hand load (LHS) is exactly balanced by the right hand load (RHS) (Fig. 9).

7 Fundamental laws of numbers

There are five fundamental laws of numbers, some, if not all, of which the reader may have used without a moment's thought! They, nevertheless, need to be stated.

Law 1 $\qquad \boldsymbol{a+b=b+a}$

Example: Suppose $a = 2$, $b = 3$ (as a test),

then $\qquad \text{LHS} = 2+3 = 5;\ \text{RHS} = 3+2 = 5$

$\therefore \qquad \text{LHS} = \text{RHS}$

Law 2 $\qquad \boldsymbol{ab=ba}$

Example: Suppose $a = 3$, $b = 5$,

then $\qquad \text{LHS} = 3 \times 5 = 15;\ \text{RHS} = 5 \times 3 = 15$

$\therefore \qquad \text{LHS} = \text{RHS}$

These two laws are the *commutative laws of addition* and *multiplication*, respectively. They state that the order of the elements may be interchanged without altering the result, when adding (law 1), or when multiplying (law 2).

Law 3 $\qquad \boldsymbol{a+(b+c)=(a+b)+c}$

Example: let $a = 3$, $b = 4$, $c = 8$, then

$$\text{LHS} = 3+(4+8) = 3+12 = 15$$
$$\text{RHS} = (3+4)+8 = 7+8 = 15$$
$$\therefore \quad \text{LHS} = \text{RHS}$$

This is the *associative law of addition*. It states that it does not matter in what order we add up the elements.

Law 4 $\qquad a(bc) = (ab)c$

Example: let $a = 6$, $b = 9$, $c = 2$;

$$\begin{aligned} \text{LHS} &= 6 \times (9 \times 2) = 6 \times 18 = 108 \\ \text{RHS} &= (6 \times 9) \times 2 = 54 \times 2 = 108 \\ \therefore \quad \text{LHS} &= \text{RHS} \end{aligned}$$

This is the *associative law of multiplication.* It states that it does not matter in what order we multiply the elements together.

Law 5 $\qquad a(b + c) = ab + ac$

This is the *distributive law*, which states that when the *sum* of two elements is multiplied by a third element the result is the same as that obtained by multiplying each of the first two elements, *in turn*, by the third element and then adding. Law 5 has already been studied in Section 6 above; the letters used were different, but this does not matter.

These five laws are valid for *any* number of elements, although they are only given above for the *minimum* numbers of elements, *two* in the case of laws 1 and 2, and *three* in the case of laws 3, 4 and 5.

Note that the numerical examples under the laws are simple verifications in particular cases – they are *not* proofs.

Extensions of Laws 3, 4 and 5 (These extensions, which are 'built in', may not be obvious to the reader.)

Law 3 $a + (b + c) = (a + b) + c = a + b + c = (a + c) + b$, etc. (using law 1)
Law 4 $a(bc) = (ab)c = abc = acb = a(cb)$, etc. (using law 2)
Law 5 $a(b + c) = ab + ac = ac + ab = a(c + b) = (b + c)a$, etc.

More elaborate extensions of the five laws will be given later, e.g. if we put $a = x + y$ in the statement of law 5 we get:

$$\begin{aligned} (x + y)\,(b + c) = (x + y)b + (x + y)c &= b(x + y) + c(x + y) \\ &= bx + by + cx + cy. \end{aligned}$$

Example 9 Solve the equation $4(\frac{1}{2}x + 3) = 15$.

We have $\qquad 4 \times \frac{1}{2}x = 2x$ and $4 \times 3 = 12$ (law 5).

$\therefore$ the equation becomes

$$2x + 12 = 15$$

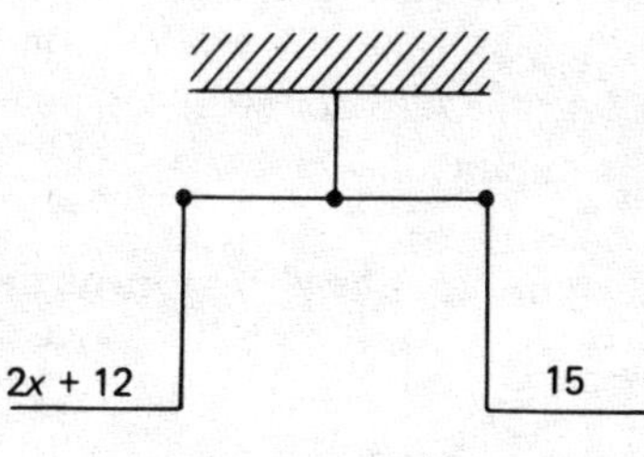

As on pages 44–5, the two sides balance; if we add, or subtract, the same amount from both sides they will still balance.

$\therefore$ on taking 12 from each side

$$2x = 3 \Rightarrow x = \frac{3}{2} = \mathbf{1.5}$$

Example 10 Solve the equation $3(3x+4)+5(x-2) = 30$.

We have $$9x + 12 + 5x - 10 = 30$$

(law 5 used twice)

Collecting together (a) the terms in x, (b) the other terms, in the LHS, we get

$$14x + 2 = 30$$

$\therefore$ on taking 2 from each side

$$14x = 28$$

$$\Rightarrow \quad x = \frac{28}{14} = \mathbf{2}$$

Example 11 Solve the equation $\frac{1}{2}(2a+3) - \frac{3}{5}a = 2$.
We firstly multiply throughout by the LCM of the denominators, i.e. $2 \times 5 = 10$.

Then $$5(2a+3) - 6a = 20$$

$\therefore$ $$10a + 15 - 6a = 20$$

$\therefore$ $$4a + 15 = 20$$

so $$4a = 5 \Rightarrow a = \frac{5}{4} = \mathbf{1.25}$$

Example 12 A party of men hires a minibus for an outing, and each man agrees to pay £2 for his part of the hire charge. There are, however, three empty seats which are later filled, and the cost per man is thereby reduced to £1.60. How many men are there in the total party and what is the hire charge for the minibus?

Let the initial number of men be x, then the final number is $x+3$. $\therefore$ Initial total cost was £$2x$ and the final total cost is £$1.6(x+3)$. (*Note*: the zero in 1.60 makes no difference here.) But the cost of hiring the minibus has not changed,

$\therefore$ (omitting the £ sign throughout)

$$2x = 1.6(x+3)$$

$$\therefore \quad 2x = 1.6x + 4.8$$

which gives $\quad 0.4x = 4.8 \Rightarrow 4x = 48 \Rightarrow x = 12$

$\therefore$ the total number of men is $x+3 = \mathbf{15}$; the hire charge is £$2x$ = £**24**.

(*Note:* The hire charge would have been the same had we taken £$1.6(x+3)$ = £1.6×15 = £24.)

The thoughtful reader may, at this stage, have asked himself why brackets should be necessary. The reason is quite easily demonstrated. Suppose we compare $3 \times p + 2q$ and $3(p+2q)$.

(a) $3 \times p + 2q = 3p + 2q$, i.e. 3 times p is added to 2 times q (multiplication precedes addition).

(b) $3(p+2q) = 3p + 3 \times 2q = 3p + 6q$ (distributive law), i.e. 3 times p plus 6 times q, which is not the same as in (a) above.

As a matter of fact our priorities are included in the mnemonic *BODMAS*, provided we remember when we may, and when we may not, simplify the contents of brackets by first collecting together some, or all, of the terms therein.

EXERCISE 2

1 Verify the following by substituting the different numerical values given, in each side of the equation concerned:

(a) $2x + 3y = 3y + 2x$, in the case when $x = 4$, $y = 7$;

(b) $pq = qp$, when $p = 2.4$, $q = 0.725$;

(c) $\frac{2}{3}a + (b + 3c) = (\frac{2}{3}a + b) + 3c$, when $a = 6$, $b = \frac{1}{2}$, $c = \frac{3}{2}$;

(d) $f(gh) = (fg)h$ when $f = 1\frac{1}{2}$, $g = \frac{4}{5}$, $h = 2\frac{2}{3}$;

(*Hint:* Firstly express the numerical values of f and h as improper fractions.)

(e) $\frac{3}{5}u(20v + 6w) = 12uv + \frac{18}{5}uw$, when $u = 6$, $v = \frac{3}{4}$, $w = \frac{1}{3}$.

2 Solve the following equations

(a) $3(x+7)=36$ (b) $4(a-2)+11=23$
(c) $\frac{1}{2}(4y-3)+3y+5=6$ (d) $\frac{2}{3}(2x+6)+\frac{1}{4}(x-12)=20$.

3 Smith buys x postage stamps at 8p each and y stamps at 10p each. (a) Give an expression for the total cost in pounds. (b) What was the *average* amount Smith paid, in pence, for a stamp?

4 The total cost of 3 lb of tea and 4 lb of sugar is 70p more than the total cost of 2 lb of tea and 7 lb of sugar. If tea is 8 times as expensive as sugar, find the cost of 1 lb of each. (*Hint*: If the cost of sugar is x pence a pound, the cost of tea is $8x$ pence a pound.)

5 Jones is four times as old as his son, David. In six years time, David will be one-third as old as his father. What are the present ages of father and son?

6 A rectangular room is of length $8c$ feet. The area of the room is $12cd$ sq. ft. Find (i) the breadth of the room, (ii) the perimeter of the room (i.e. the distance all round the edge – usually taken around the floor).

4

Some Ideas of Geometrical Reasoning

1 Introduction and a few definitions

We are now going to take a look at some of the interesting things we can do with points, lines, angles and simple geometrical shapes. Some of our earlier algebraic ideas will be applied to them, in order to avoid the long and formal presentation required by strictly Euclidean methods. Firstly, however, we need some definitions, *which at present we are restricting to a plane* (a flat surface).

1 Elementary geometry is that section of mathematics which deals with the sizes and shapes of things; we shall, however, use a slightly different definition implied by the paragraph above, namely, that geometry is concerned with the properties of points, lines, surfaces and solids.

2 A point To most people, a point is a dot, but the following definition is rather more precise. A point is a geometrical element which has position but no magnitude, i.e. its size is zero. This is very inconvenient, so we do put dots to indicate points, but we make sure the dots are small!

3 A line This can be defined in two ways: (a) a line is a path traced out by a moving point, (b) a line is that which has length but no breadth. Putting these together, we have a pretty clear idea of what is meant. A line may be curved or straight and, on a diagram, should be thinly drawn in pencil or ink. *Note: Unless otherwise stated, from now on, our use of the word 'line' will be taken to mean 'straight line'.*

4 *A straight line* The popular definition of a straight line is that of being the shortest distance between two points. This presupposes that we cannot draw a line without predetermining two points of it, which need not be the case at all. There seems, in fact, to be no fully satisfactory definition, but fortunately almost everyone understands what is meant when talking about a straight line. A good exposition by James and James, *Mathematics Dictionary* (1949) (D. Van Nostrand Co. Inc.) is that a straight line is 'A line such that if any part of it is placed so as to have two points in common with any other part, it will lie along the other part'.

5 *An angle* (a) An angle can be a corner of a figure – a point from which lines radiate; (b) an angle can be the inclination of one line to another. The different meanings are worthy of a moment's careful thought: (a) refers to *part* of a *figure*, whereas (b) considers the *size* of the angle.

Fig. 10 illustrates the above definitions (2–5); in 5b(i), KG is a line, and in 5b(ii) RS and PQ are intersecting lines.

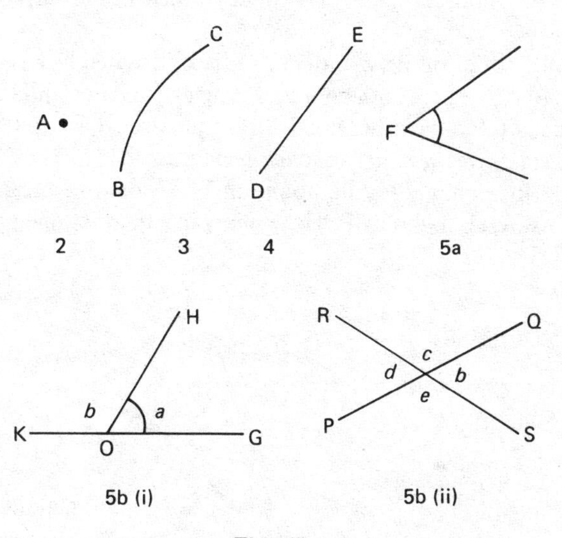

Fig. 10

In Fig. 10, diagrams 2, 3, 4 and 5a are easily understood; diagram 5b(i), in addition to $\hat{a}$ (i.e. $G\hat{O}H$) has $\hat{b}$ (i.e. $H\hat{O}K$); diagram 5b(ii) has $\hat{b}$, $\hat{c}$, $\hat{d}$

and $\hat{e}$ between the lines PQ and RS, intersecting at T. (For the explanation of ^ , see Section 2 below.)

2 Angles

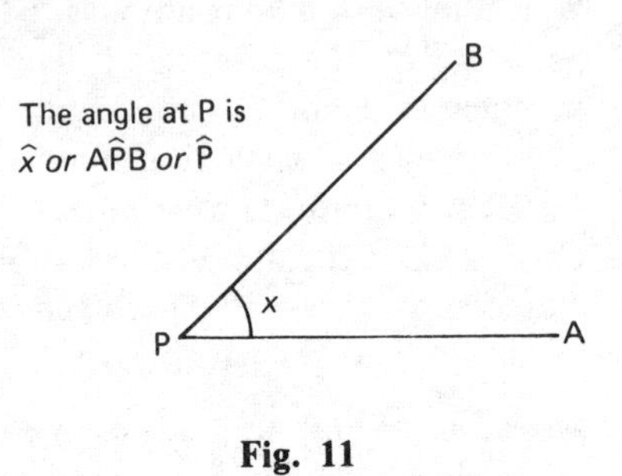

Fig. 11

These, as already shown, may be represented by one lower case letter lying within the *arms* PA, PB of the angle, e.g. $\hat{x}$ in Fig. 11 (N.B. when this letter is written, *other than in the diagram, it must have an angle symbol*; we shall use a circumflex accent over the letter). Alternatively, it may be given as A$\hat{\text{P}}$B (Fig. 11). If, however, only two lines radiate from the corner (as in Fig. 11) we may use the single capital letter at that corner.

Definition 6 *Adjacent angles* are angles which lie side by side at a point.

In Fig. 12, for example, $\hat{p}$ and $\hat{q}$ are adjacent. When however AP is produced to A′, we get another angle, $\hat{r}$. Angles $\hat{q}$ and $\hat{r}$ are adjacent, but $\hat{p}$ and $\hat{r}$ are not. On the other hand A$\hat{\text{P}}$C and $\hat{r}$ are adjacent. It follows that *adjacent angles need to have one common arm and to lie on opposite sides of it.* (Incidentally to talk about angle P would be meaningless. There are several angles at P. How many are there altogether?)

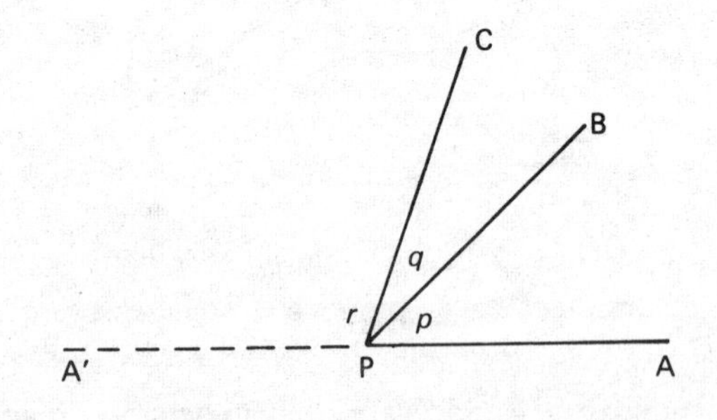

Fig. 12

Definition 7 Vertically opposite angles When two straight lines AOA′, BOB′ intersect at O (Fig. 13), the angles AOB and A′OB′ are *vertically opposite* and, likewise, so are angles BOA′ and B′OA.

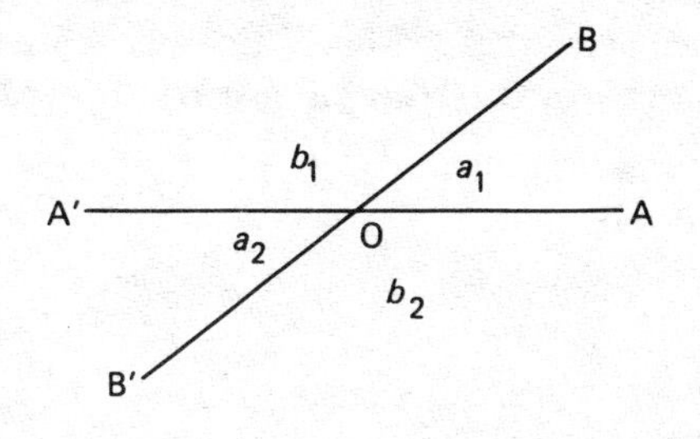

Fig. 13

We now need our first *axiom* (or *postulate*). An axiom is a generally accepted principle or 'self-evident' truth. It is not something which permits of direct proof.

Axiom 1 Vertically opposite angles are equal. Thus, in Fig. 13:

$$\hat{a}_1 = \hat{a}_2 \text{ and } \hat{b}_1 = \hat{b}_2$$

The lettering was deliberately chosen. If we know two angles are equal, by using the same letter for them, and at the same time distinguishing them by using different suffixes such as 1 and 2, we visualise the picture clearly. There are occasions where, however, it may be better to use separate letters for equal angles.

3 The measurement of angles

The measurement of time by a clock starts at the top of the dial and rotates *clockwise* (reasonably enough!) through a full circle, which completes one hour of 60 minutes. Likewise, in navigation we start at the top (North) on a chart and move *clockwise* through a complete circle, which is made up of 360 degrees, written 360°; each degree is divided into 60 minutes of arc, written 60′. The measurement of time and of angle (in navigation and surveying) are both based on the Babylonian sexagesimal scale, mentioned in Chapter 1.

The traditional method of representing course or bearing at sea *was* from North to 90° East or West, or from South to 90° East or West. This method is still used in some small craft. The modern system, especially in ships provided with gyrocompasses, is to measure from 000° (North), clockwise through 090° (East), 180°(South), 270° (West) up to 360° (which is then read as 000°). Compass-cards, including, by the way, those used in orienteering, often carry both scales, traditional and modern (Fig. 14).

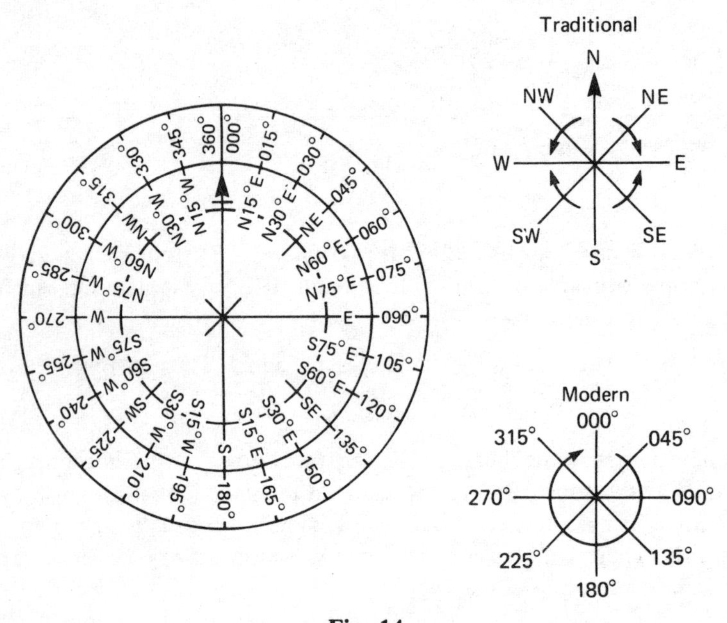

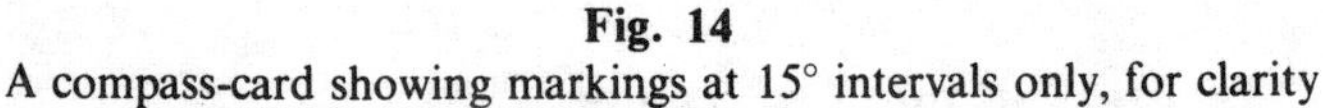
Fig. 14
A compass-card showing markings at 15° intervals only, for clarity

The following example illustrates the *use* of the 000° to 360° compass card in a ship equipped with a *gyrocompass*, which has been set without gyro error. This avoids any necessity for correction of variation and/or deviation, present when a magnetic compass is used.

Example 1 A ship leaves Dover at 0740 hours, sailing on a course of 138° (true) at a speed of 12 knots (nautical miles per hour) until 0830, when she alters course to 094° and simultaneously increases speed to 14 knots. Find, by scale drawing using a ruler and a protractor, her bearing and distance from Dover harbour at 0900, giving the result correct to the nearest degree and 0.2 nautical mile, respectively.

(*Notes*: Firstly, read Chapter 6, Section 3, which largely concerns the protractor. Secondly, use graph paper, if available, as this makes the measurement of angles by protractor considerably easier to carry out.)

Let D be Dover, the starting point at 0740; let E be the position at 0830, and F the position at 0900.

Due North, marked N on the diagram, is in the direction of the top of the paper from whatever points we are measuring (in this case, D and E).

Mark off $N\hat{D}E = 138°$ at D.

The time from D to E is from 0740 to 0830, i.e. 50 min, ∴ the distance $DE = \frac{50}{60} \times 12$ n.m. = 10 n.m. (*n*autical *m*iles)

Lay off DE = 10 n.m. (to a scale, say, of 1 cm to 2 n.m).

Mark off $N\hat{E}F = 94°$ at E.

The time from E to F is from 0830 to 0900, i.e. 30 min, the distance $EF = \frac{30}{60} \times 14$ n.m. = 7 n.m.

Lay off EF = 7 n.m. to the same scale as previously used.

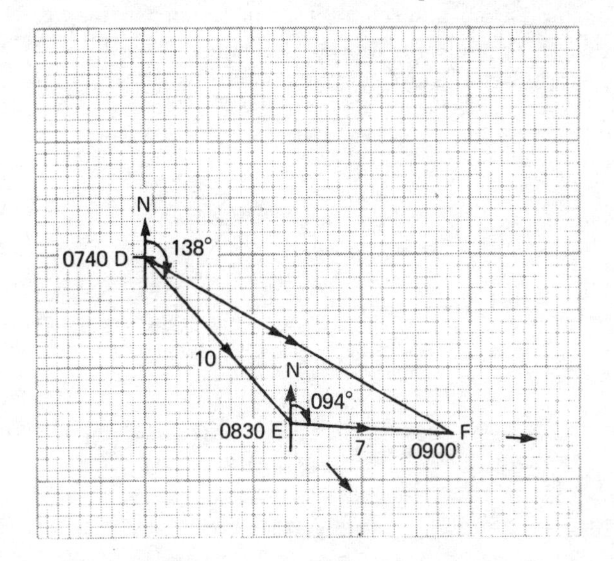

Join DF. This gives the bearing (N$\hat{D}$F, measured at D) and the distance (DF) of the ship from Dover harbour. By measurement:
(i) the bearing is **120°**, (ii) the distance is **15.8 nautical miles.**

Those readers who have had little experience of the sea may like to refer back for a moment to the compass-card, Fig. 14. North, East, South and West are called the *cardinal points.* This brings to mind that competent seamen until recent times had to 'box the compass', i.e. to know the 32 *points* of the compass, to state them in correct order (clockwise), and then reverse order – faultlessly! The points ran at $11\frac{1}{4}°$ intervals, starting at North, thus: N, N by E, NNE, NE by N, NE, NE by E, ENE, E by N, E, E by S, ESE, and so on.

There were eight points ($8 \times 11\frac{1}{4}° = 90°$) from one cardinal point to the next, and hence 32 for a complete rotation.

Returning to *mathematical* convention, we again take 360° for a full revolution, but we generally rotate in an *anticlockwise* direction. We sometimes start from the right, i.e. from the direction marked East in Fig. 14, although this is not always convenient.

Fig. 15 illustrates the convention often adopted. It will be observed that in mathematics we do *not* need to put in the initial zeros for angles of less than 100° (as we do in gyro-navigation), e.g. Mathematics 72°, Navigation 072°. This is because, in navigation, courses and bearings are measured from a specific direction (often North: e.g. N 20°E, N 47°W, 062°; and sometimes South: e.g. S 44°E, S 59°W). In mathematics, we take any convenient starting direction.

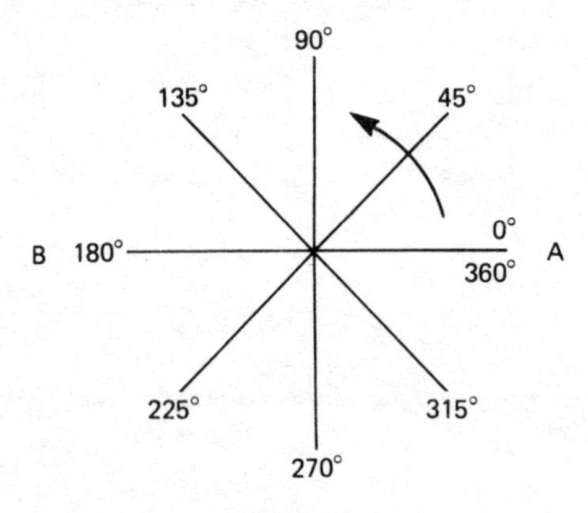

Fig. 15

Fig. 16 shows that a rotation of 180° from A takes us to B on the line AOB, and AOB can be called a *straight angle*. This stratagem is sometimes useful.

It will be recalled that the definition of adjacent angles above indicates that they lie side by side and have a common arm between them. If we draw a line AOB, and at O erect a line OC such that $A\hat{O}C = C\hat{O}B$ (Fig. 16), i.e. $\hat{a} = \hat{b}$, then $\hat{a}$ and $\hat{b}$ are *right angles*, i.e. each is one half of a straight angle, which is itself one half of a complete rotation (360°).

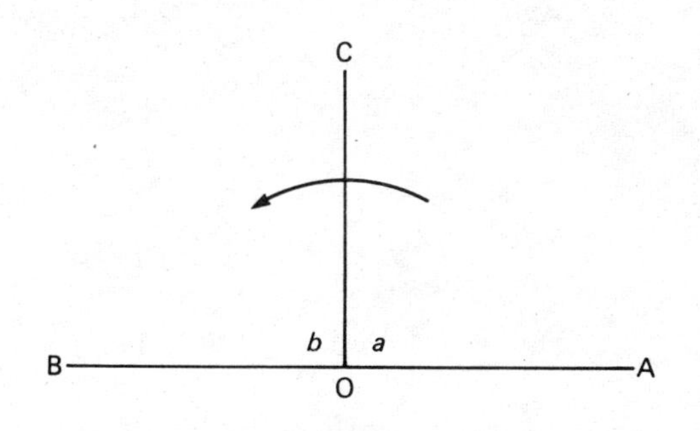

Fig. 16

Proof. We know that $A\hat{O}B = 180°$, $\therefore\ \hat{a}+\hat{b} = 180°$, but $\hat{a} = \hat{b}$, by our construction.

$$\therefore \qquad 2\hat{a} = 180° \quad \text{i.e. } \hat{a} = 90° = \hat{b}$$

That is, *a right angle is an angle of* 90°. A right angle is shown on a diagram as ∟ in the angle itself. Furthermore, OC is said to be *perpendicular* (i.e. at right angles) to AB.

Definitions

8 An *acute angle* is one which is less than 90° (i.e. *less* than *one* right angle).

9 An *obtuse angle* is one which lies between 90° and 180° (i.e. between *one* and *two* right angles).

10 A *reflex angle* is greater than 180° and less than 360° (i.e. between *two* and *four* right angles).

These are illustrated in Fig. 17.

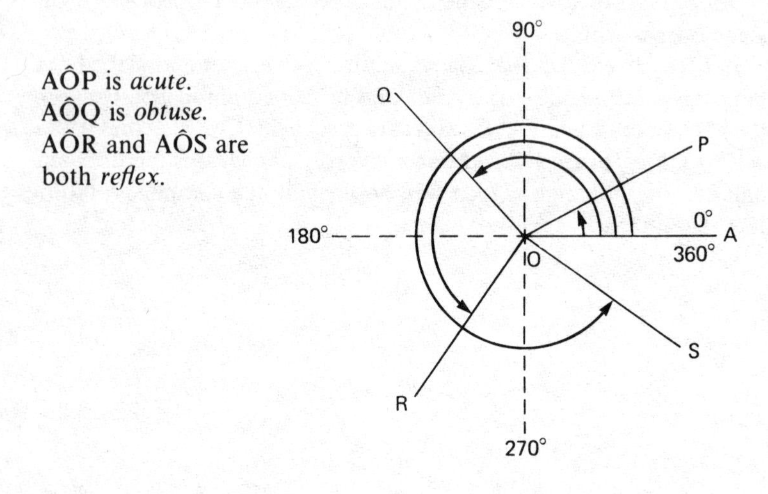

Fig. 17

11 Complementary angles are adjacent angles whose sum is a right angle (90°).

12 Supplementary angles are adjacent angles whose sum is a straight angle (180°).

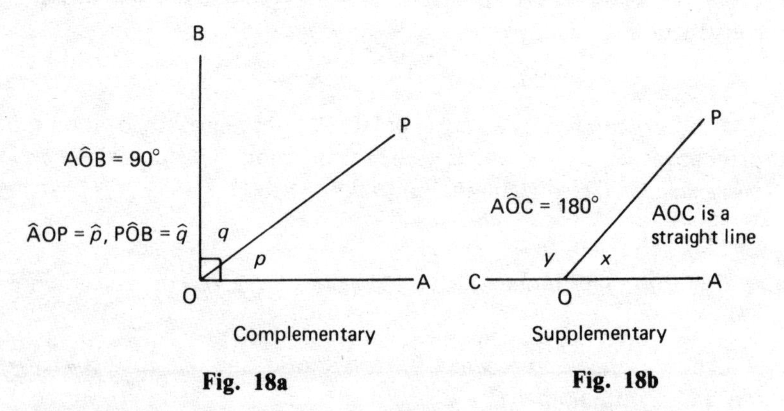

Fig. 18a **Fig. 18b**

In Fig. 18a, $\hat{p}+\hat{q} = 90°$, i.e. $\hat{p}$ and $\hat{q}$ are complementary.
In Fig. 18b, $\hat{x}+\hat{y} = 180°$, i.e. x and y are supplementary.

It is worth noting, in passing, that angles can go on increasing beyond 360°. The hands of a clock pass through 60 minutes and then repeat the path for the following hour. Likewise the pencil in a pair of compasses can be rotated, indefinitely retracing a circle once it has been completed. In so far as the final position of a rotating arm relative to its starting position is concerned, the visual angle turned through is merely the total angle of rotation *less* the largest whole-number (integral) multiple of 360° which will leave the final answer positive and less than 360°. For example, consider 520° and 1170°.
(i) $520° - 360° = 160°$, (ii) $1170° - 3 \times 360° = 1170° - 1080° = 90°$

At this stage we need to understand the concept of a *theorem*, which is a principle or general statement which is accepted, established or demonstrable (i.e. it can be proved). The methods of geometrical proof are interesting and, in some cases, fascinating, particularly for people who enjoy logical reasoning in games and puzzles. Space, alas, does not here permit more than the use of some simple theorems; and, likewise, proofs will be few and far between!

4 Parallel lines

Definition 13 *Parallel lines* are lines which are equidistant from one another, i.e. no matter how far they are produced, they will not meet at a point. *Lines which are parallel are often drawn with arrows* on them, but this convention is *not* the same as was used in the earlier navigation example.

There are some useful properties associated with parallel lines, given in the theorem below.

Theorem 1 When a line cuts two parallel lines (i) the *alternate angles* are equal, (ii) the *corresponding angles* are equal, (iii) the *co-interior angles* are supplementary.

Fig. 19 shows examples (i), (ii) and (iii).

All three diagrams in Fig. 19 represent the same parallel lines AB and CD, cut by the same line XY (which is called a *transversal*).

There exist considerably more relationships about the angles at the points of intersection, which we now call E and F, than are shown in Fig. 19 (see Fig. 20). All the $\hat{a}$'s are equal; all the $\hat{b}$'s are equal; also $\hat{a} + \hat{b} = 180°$, for any one of the $\hat{a}$'s and any one of the $\hat{b}$'s.

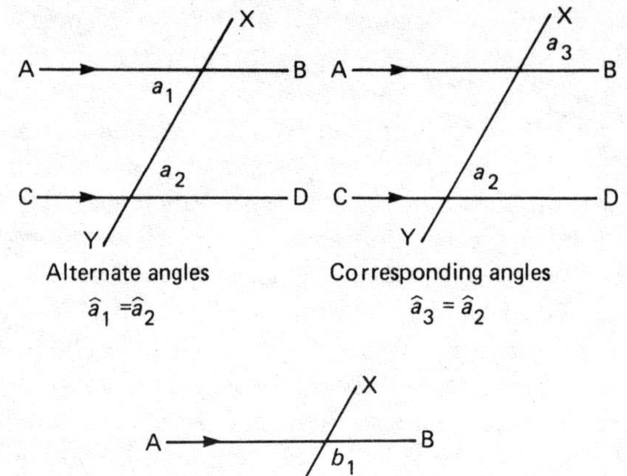

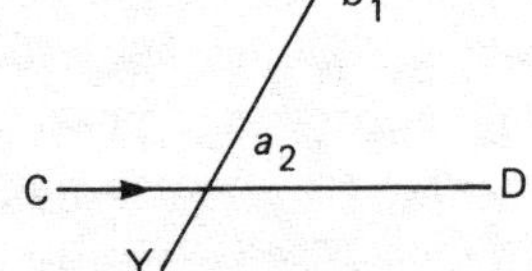

Co-interior angles
$\hat{a}_2 + \hat{b}_1 = 180°$ (supplementary)

Fig. 19

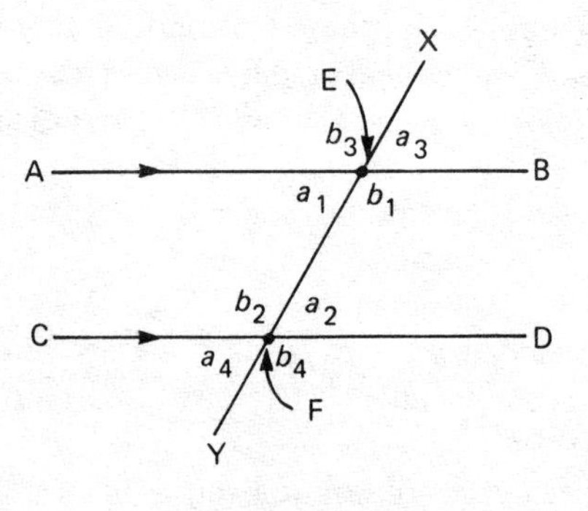

Fig. 20

The distance between two parallel lines

Let us take two parallel lines PQ and RS. We take a point T on PQ and draw TU perpendicular (i.e. at right angles) to RS, meeting it at U. Then we say that TU is the distance between PQ and RS (Fig. 21).

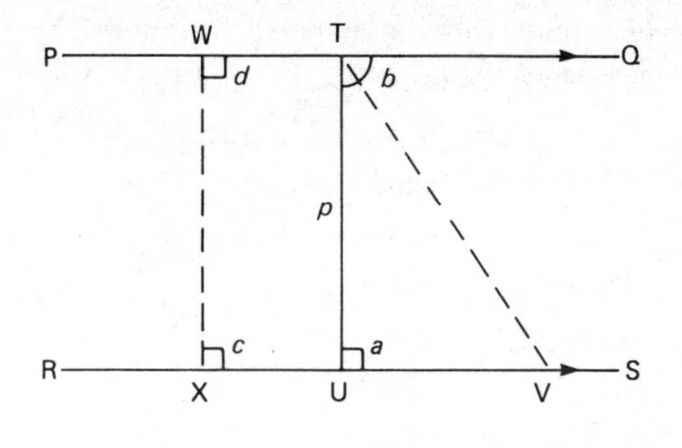

Fig. 21

From our construction $\hat{a} = T\hat{U}S = 90°$, but $\hat{a} + \hat{b} = 180°$ (co-interior angles are supplementary).

$$\therefore \qquad \hat{b} = 180° - \hat{a} = 180° - 90° = 90°$$

i.e. the line TU is a *mutual perpendicular* to RS and to PQ. If TU is of length p, then no matter where on PQ (e.g. at W) we draw a line perpendicular to RS meeting it at X, we have $\hat{c} = \hat{d} = 90°$ and $WX = p$. Thus, the distance between two parallel lines is constant. Also, if we draw any other line from T, say TV, which is *not* perpendicular to RS, then TV > TU, i.e. TV *is greater than p*. (The proof is omitted.) Incidentally, TUXW is a rectangle.

5 **Quadrilaterals** (see Fig. 22)

Definitions

14 A *quadrilateral* is a four-sided figure (which we shall restrict to the case where no side crosses over another side).

15 A *trapezium* (plural: trapezia or trapeziums) is a quadrilateral with one pair of parallel sides.

16 A *parallelogram* is a quadrilateral with both pairs of opposite sides parallel. (See deducible properties, below.)

17a A *rectangle* is a parallelogram with one angle a right angle. (See deducible properties.)

17b A *rhombus* (plural: rhombi) is a parallelogram with two adjacent sides equal. (See deducible properties.)

18 A *square* is a rectangle with two adjacent sides equal. (See deducible properties.)

A family tree

Each quadrilateral is derived from the preceding one by adding an extra condition (Fig. 22).

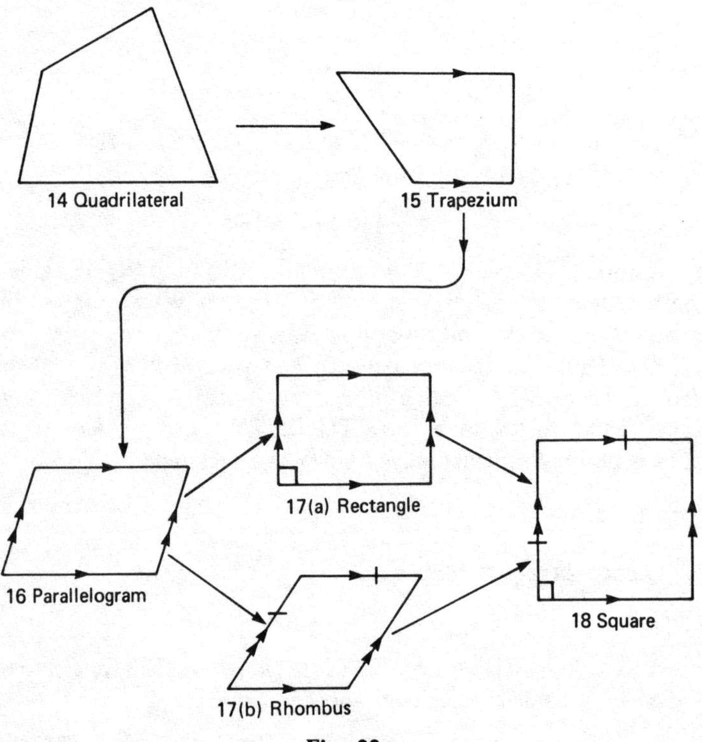

Fig. 22

These figures illustrate the *minimal* definitions; they do not show all the derivations given below.

Deducible properties of certain quadrilaterals

The proofs are all simple, and one or two brief references are given to them for use after reading Section 1 of Chapter 6. The reader would, however, do well to learn the properties themselves. They can be useful in real life.

In a *parallelogram* (definition 16), (a) the opposite sides are equal, (b) the opposite angles are equal, (c) the diagonals bisect each other. (A *diagonal* of a quadrilateral of any shape is a line joining two opposite vertices; to *bisect* a line or angle is to divide it into two *equal* parts.)

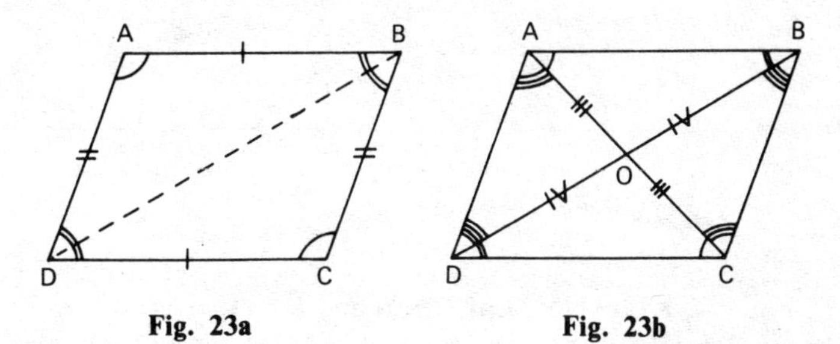

Fig. 23a **Fig. 23b**

In Fig. 23a, AB = DC and AD = BC; also, $\hat{A} = \hat{C}$ and $\hat{B} = \hat{D}$. (*Outline of proof*: Join BD, then $\triangle$ ABD $\equiv$ $\triangle$ CDB (SAS); see Chapter 6; leave consideration of the proof until that chapter has been read.)

In Fig. 23b, the same parallelogram, AO = OC and BO = OD. (*The proof* continues from that for Fig. 21a, by afterwards showing that $\triangle$ AOB $\equiv$ $\triangle$ COD (AA Cor. S); same comment as above applies.)

In a *rhombus* (definition 17b), (a) all four sides are equal, (b) the opposite angles are equal, (c) the diagonals bisect each other at right angles. Figs 24a and 24b represent the same rhombus.

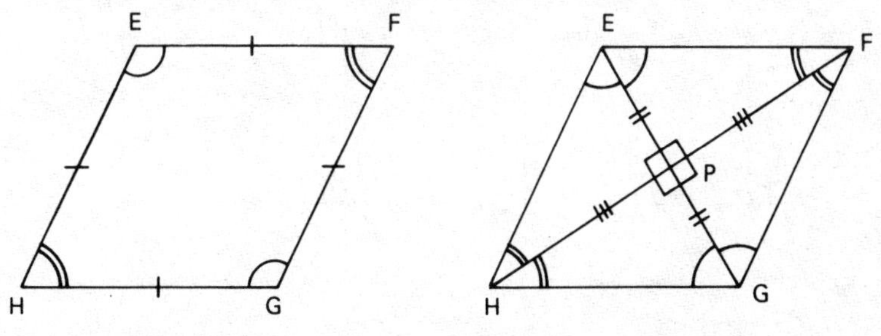

Fig. 24a **Fig. 24b**

In Fig. 24a, EF = FG = GH = HE, $\hat{E} = \hat{G}$, $\hat{F} = \hat{H}$.

In Fig. 24b, EP = PG, FP = HP; the angles marked at P are all 90°.

In a *rectangle* (definition 17a), (a) the opposite sides are equal, (b) the angles are all right angles, (c) the diagonals are equal and bisect each other. Figs 25a and 25b represent the same rectangle.

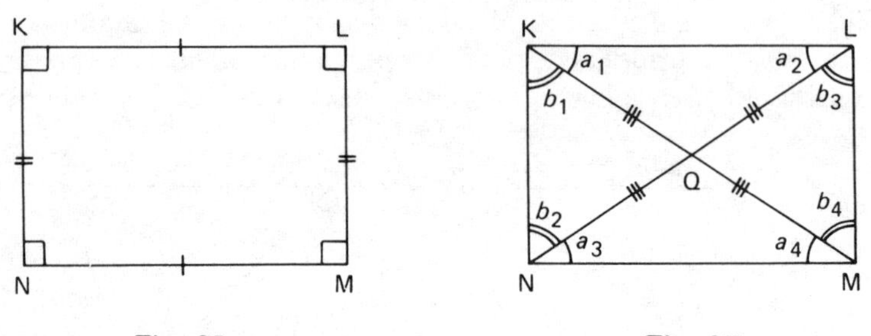

Fig. 25a **Fig. 25b**

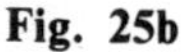

In Fig. 25a, KL = NM, LM = KN and $\hat{K} = \hat{L} = \hat{M} = \hat{N} = 90°$.

In Fig. 25b, KQ = LQ = MQ = NQ and KM = LN; all the angles marked *a* are equal, as also are all angles marked *b*,

i.e. $a_1 = a_2 = a_3 = a_4$ and $b_1 = b_2 = b_3 = b_4$.

In a *square*, (a) all the sides are equal, (b) all the angles are equal, (c) the diagonals are equal and bisect each other at right angles. (In fact, the square is a rectangular rhombus, and incorporates both sets of properties!) Figs 26a and 26b represent the same square.

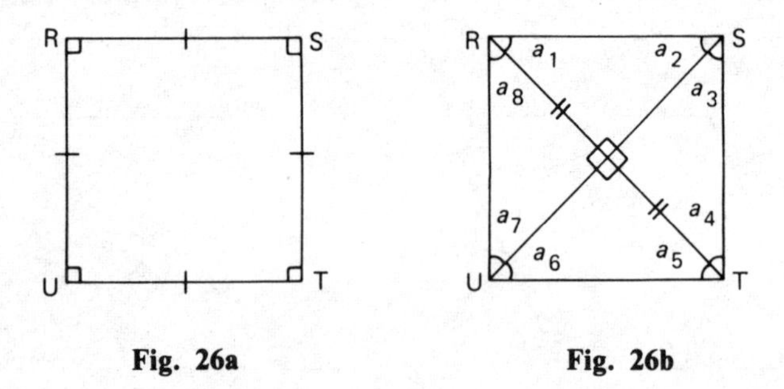

Fig. 26a **Fig. 26b**

In Fig. 26a, RS = ST = TU = UR and angles R, S, T, U are all 90°.

In Fig. 26b, RV = SV = TV = UV and the angles marked at V are all 90°; all the angles $a_1, a_2 \ldots a_8$ are equal.

Example 1 In the given diagram, find the value of the obtuse angle PQR.

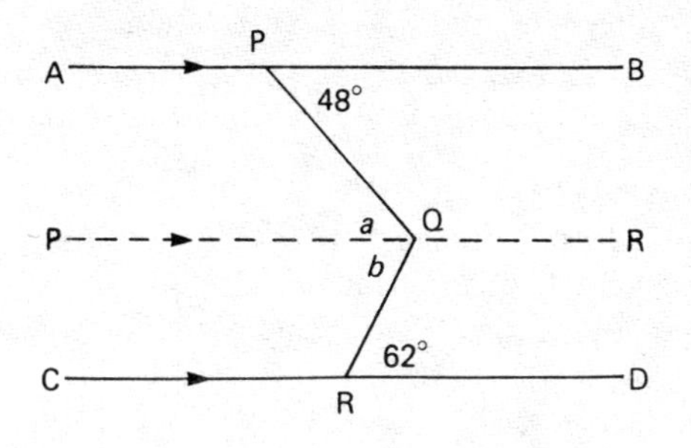

We draw a line PQR through Q, parallel to AB (and CD). It is shown dotted in the diagram.

Then

$\hat{a} = Q\hat{P}B$	(alternate angles)	‖	$\hat{b} = Q\hat{R}D$	(alternate angles)
$= 48°$	(given)	‖	$= 62°$	(given)

$\therefore$ $\quad$ $PQR = \hat{a} + \hat{b} = 48° + 62° = \mathbf{110°}$

(Note that reflex $P\hat{Q}R$, which we did not require, is $360° - 110° = 250°$.)

Definition 19 A *triangle* is a figure formed by connecting three *non-collinear* points (i.e. ones which are not in the same straight line) by lines which terminate at these points. The points themselves are the vertices of the triangle.

Example 2 (*Theorem* 2) Prove that the sum of the three angles of any triangle is 180°.

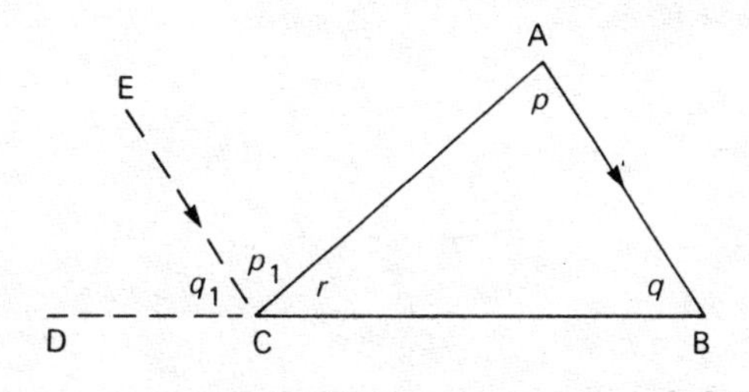

Produce BC to D, and draw CE || BA. Let the angles at A, B, C be $\hat{p}, \hat{q}, \hat{r}$ respectively; put ECA = $\hat{p}_1$ and ECD = $\hat{q}_1$, then as CE || BA,

$$\hat{p} = \hat{p}_1 \text{ (alternate) and } \hat{q} = \hat{q}_1 \text{ (corresponding)}$$
$$\therefore \quad \hat{p}+\hat{q}+\hat{r} = \hat{p}_1+\hat{q}_1+\hat{r}$$
$$= \hat{r}+\hat{p}_1+\hat{q}_1 = B\hat{C}D = 180° \text{ (a straight angle)}$$

i.e. the three angles of triangle ABC add up to 180°.

The angle ACD formed above by producing a side of the △ABC is called an *exterior angle* of the triangle.

There is also a *corollary* (subordinate theorem) which can be derived from Theorem 2.

We see that

$$\hat{p}_1 = \hat{p} \text{ and } \hat{q}_1 = \hat{q}$$
$$\therefore \quad \hat{p}_1+\hat{q}_1 = \hat{p}+\hat{q}, \quad \text{but} \quad \hat{p}_1+\hat{q}_1 = A\hat{C}D.$$
$$\therefore \quad ACD = \hat{p}+\hat{q},$$

i.e. an exterior angle of a triangle is equal to the sum of the two interior opposite angles ($\hat{t} = \hat{p}+\hat{q}$, in our redrawn diagram alongside).

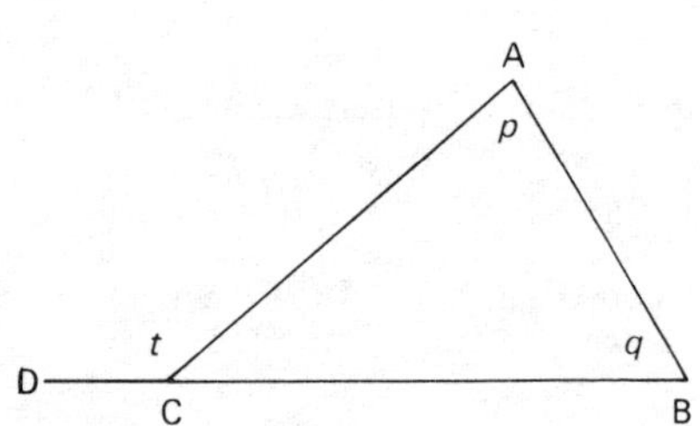

Example 2 illustrates the elegance of geometrical proofs, but *only very simple exercises of this type* will be set here for solution by the reader, as constructions and proofs generally need a more sophisticated grounding than can be provided in a book such as this.

EXERCISE 1

In each of the following figures find an equation which connects the angles named. The first is solved as an illustration; write down the results of items 2 to 8 in a similar way.

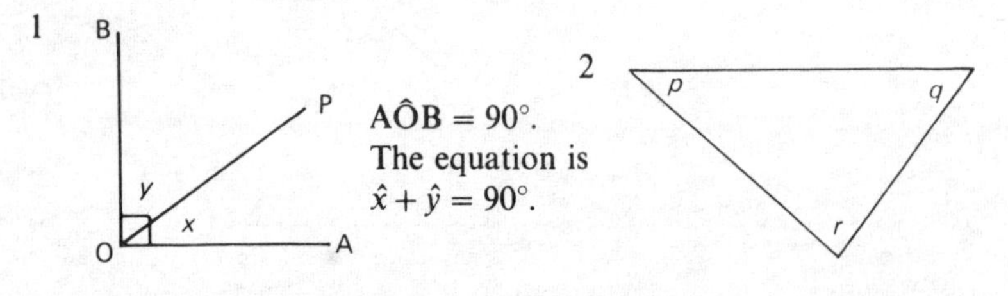

$A\hat{O}B = 90°$. The equation is $\hat{x}+\hat{y} = 90°$.

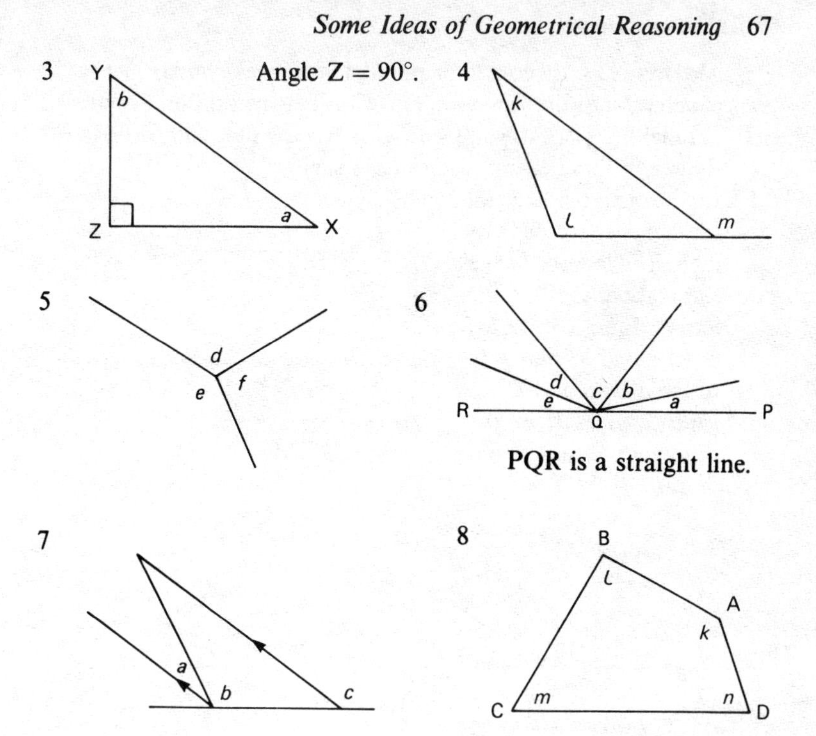

Hint: Join AC; consider the triangles.

9 In the given figure, find the values of angle ACB and reflex angle ACB.

10 In the figure ABCD, AB∥DC and BC∥AD. Show that $\hat{A} = \hat{C}$ and $\hat{B} = \hat{D}$.

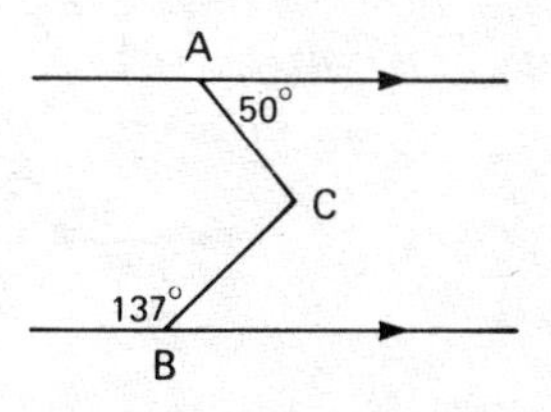

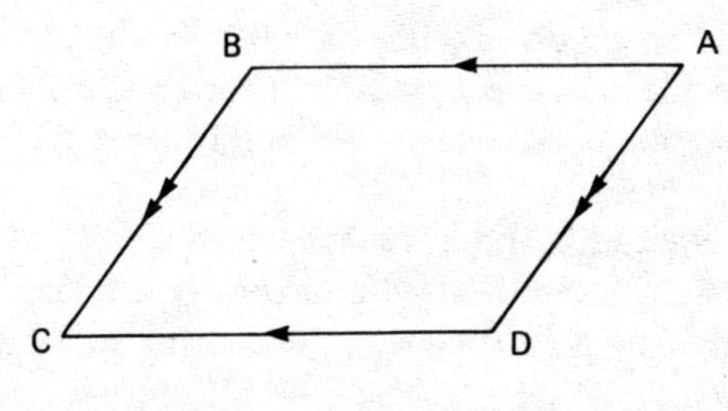

The figure is, of course, a parallelogram (as shown in Fig. 22 earlier). Note that to save confusion between different pairs of parallel lines, we use single arrows for the first pair, double for the second and so on, where necessary.

11 In the given figure, all the angles are 90° or 270°. The lengths of the *sides* are given by the letters. Find x in terms of a, c and e; find y in terms of b, d and f.

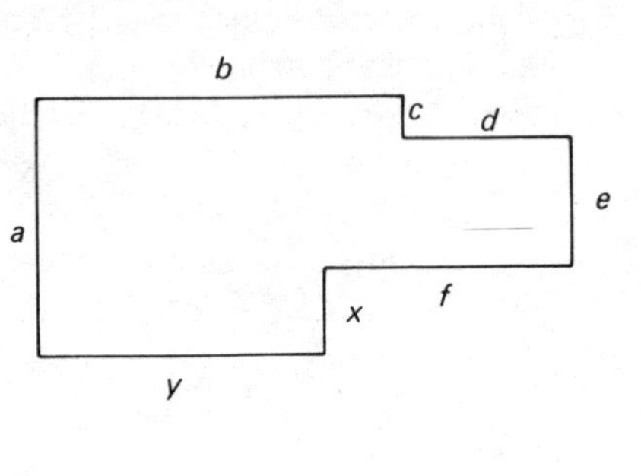

12 Find $\hat{x}$ in each of the following three diagrams:

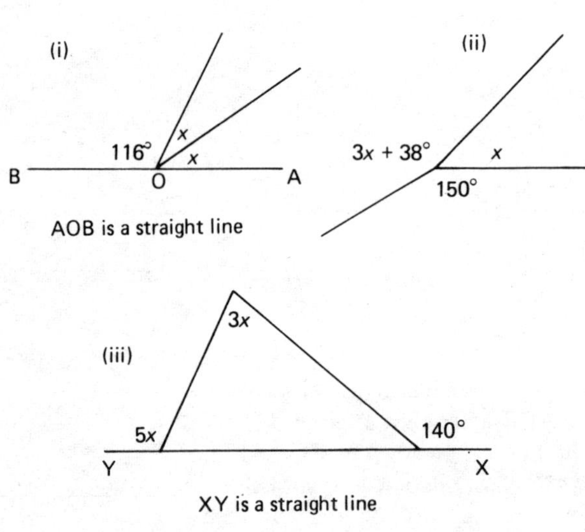

13 A ship is steering on a course of 320°. She then alters course by 90°. Find the possible new course (or courses). (*Reminder*: In navigation the *course*, when given purely in number form, always has three digits, e.g. 078°, 293°).

14 A quadrilateral has three angles each of 90°. What is the size of the fourth angle and what is the shape of the figure? (see Fig. 22.)

15 A four-sided figure is such that three of the angles are $\hat{x}$, $2\hat{x}$, and

$3\hat{x}$, respectively, where $\hat{x}$ is measured in degrees. The fourth angle is 48°. Find the other angles.

Had the fourth angle been 54° instead, what would have been the value of the smallest angle?

16 Two adjacent angles of a quadrilateral are 35° and 145°. What kind of a quadrilateral is it? (Assume that there are no other special properties, as listed in the earlier definitions and deductions, in this quadrilateral.)

17 A ship leaves harbour at 14 00 hrs and steams at a steady speed of 15 knots throughout her voyage. Initially she steers on a course of 065°. At 15 20 hrs she alters course to 098°, and at 16 56 hrs alters course again, now steering 127°. What is her bearing and distance from the harbour, at 18 40 hrs? (See sketch.) Use graph paper.

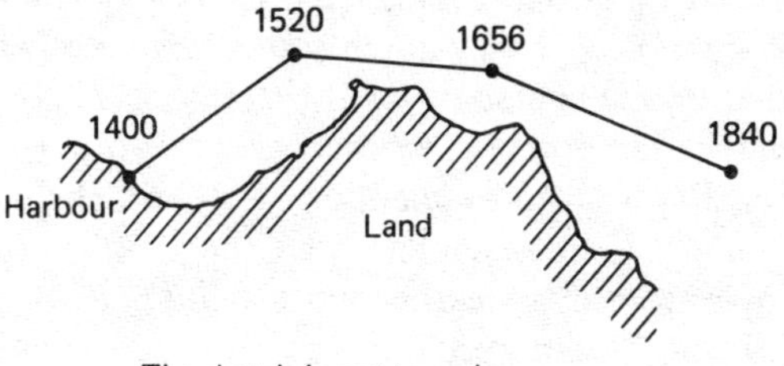

The sketch is *not* to scale

18 A ship leaves harbour A at 06 00 and proceeds on a course of 212° at 12 knots. Another ship leaves harbour B, which is 60 nautical miles due South of A, at 07 15, and steers 314° at 14 knots. How far apart will the ships be at 08 45?

What is the bearing of the second ship from the first, half an hour later still? Are the ships on collision course?

5

Directed Numbers and Extension of Indices

1 Historical note

To the uninitiated, the two topics named in the heading may appear to have little in common, but they are in some ways closely associated. One hopes that in due course this will become clear!

It is known that parentheses, i.e. brackets of the form () which have already received introductory mention, were widely used in the seventeenth century. Historical references to negative numbers, on the other hand, date from much earlier times. Although Diophantus of Alexandria (fl. *c.* 250 AD) certainly did not understand *negative* solutions to algebraic equations, we know that somewhere within the period bounded by the fifth and seventh centuries AD the Hindus were happily discussing *positive* and *negative* numbers (i.e. ones with + and − signs, respectively). If we represent positive numbers in one direction (which is usually to the right, or upwards) and negative numbers in a direction exactly opposite to the corresponding positive numbers, the numbers are said to be *directed.*

2 Directed numbers

We can represent numbers on a straight line which is called the *number line.* As indicated above, we take positive numbers as being towards the right from a starting point which is zero and is called 0, the origin. Negative numbers are measured towards the left from 0 (see Fig. 27).

A positive number is entitled to a *plus* sign, which is omitted when confusion cannot arise, e.g. number 15 may be written as 15 or +15, according to circumstances. Likewise, number a may be given as $+a$ or a (even though a itself may later have a negative value given to it). One says, for example, 'There were 15 people at the party', *not* 'There were +15 people at the party'. When, however, a number is added to another in front of it, say, 'add x and y' we put $x+y$; in such a case, the first number does not need a plus sign, but the second certainly does.

A negative number, in mathematics, always requires a *minus* sign to be placed before it, e.g. *minus* 7 is written as -7, and *minus* b as $-b$ (even though b itself may turn out later to be negative).

Not only does the number line contain zero and all the integers, i.e. $0, 1, 2, 3 \ldots$ and $-1, -2, -3 \ldots$, but all the intermediate numbers as well, e.g. 2.5, -3.8, $\sqrt{2}$ ($\simeq 1.414$), and so on.

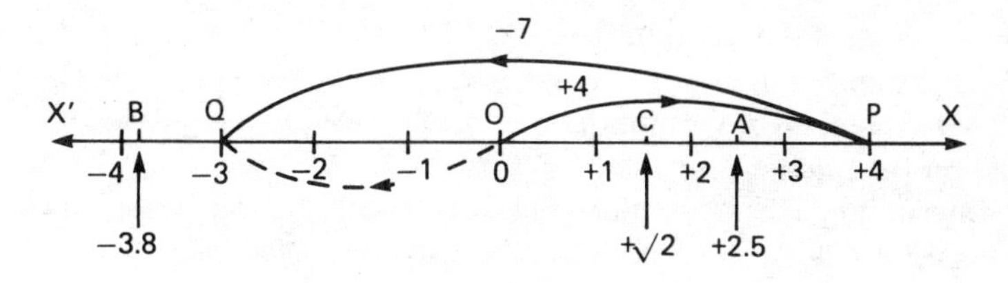

Fig. 27

The line can be extended indefinitely at both ends. (It can be considered as running from $-\infty$ to $+\infty$, where the symbol ∞ means 'infinity'). If we call the right hand 'end' X and the left hand one X′, this will be helpful later when we come to graphical work. In Fig. 27, OA = 2.5, OB = -3.8 and OC $\simeq \sqrt{2}$.

Suppose Dr Jones, who lives at O, has to visit two patients who live on opposite sides of his surgery, at P and Q respectively, one being 4 miles away and the other 3 miles. Again in Fig. 27, OP = $+4$ (miles) and OQ = -3 (miles). It is obvious that if he starts by going from O to P, a distance of $+4$ miles, he then has to travel from P to Q a distance of -7 miles (where PQ is measured in the opposite direction from OP). Now, on the number line, OP = 4, PQ = -7 and OQ = -3 and we therefore deduce that $+4-7=-3$.

Note: In moving from O to Q, it is important to distinguish between (i) the use of directed numbers (above), which give the displacement of Dr Jones (at Q) from his starting point, O, and (ii) the actual distance he travels to get to Q:

(i) The *displacement* OQ is -3 miles; if we ignore direction, this is 3 miles, numerically;
(ii) The *distance travelled*, which automatically ignores direction, is $4+7$ miles $= 11$ miles.

Another example: Robinson pays £600 into his bank account on the first day of each month. During January, February and March, he withdraws £580, £620 and £640 respectively. We compare his balances of account at the end of each month.

Let the months be represented by their initial letters.

(a) J. At the end of the month he has £(600 − 580) = £20, i.e. +20 on the number line
(b) F. At the end of the month he has £(20 + 600 − 620) = £0 i.e. 0 on the number line
(c) M. At the end of the month he has £(600 − 640) = −£40, i.e. −40 on the number line

Interpreting these results (Fig. 28), Robinson has (a) a *credit* balance of £20, (b) nothing left in the kitty, (c) a *debit* balance of £40 (i.e. he owes the bank this amount); this debit balance can be considered as a credit balance of −£40.

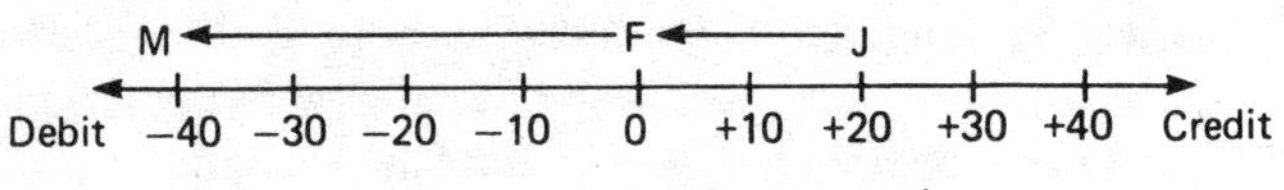

Fig. 28

3 Addition and subtraction of positive and negative numbers

Addition

There are four possible cases, which are illustrated in Fig. 29. They are equivalent to, e.g. (a) £5 added to £2 = £7, (b) a gain of £6 added to a loss of £2 = a gain of £4, (c) a loss of £4 added to a gain of £7 = a gain of £3, (d) a loss of £2 added to a loss of £4 = a loss of £6.

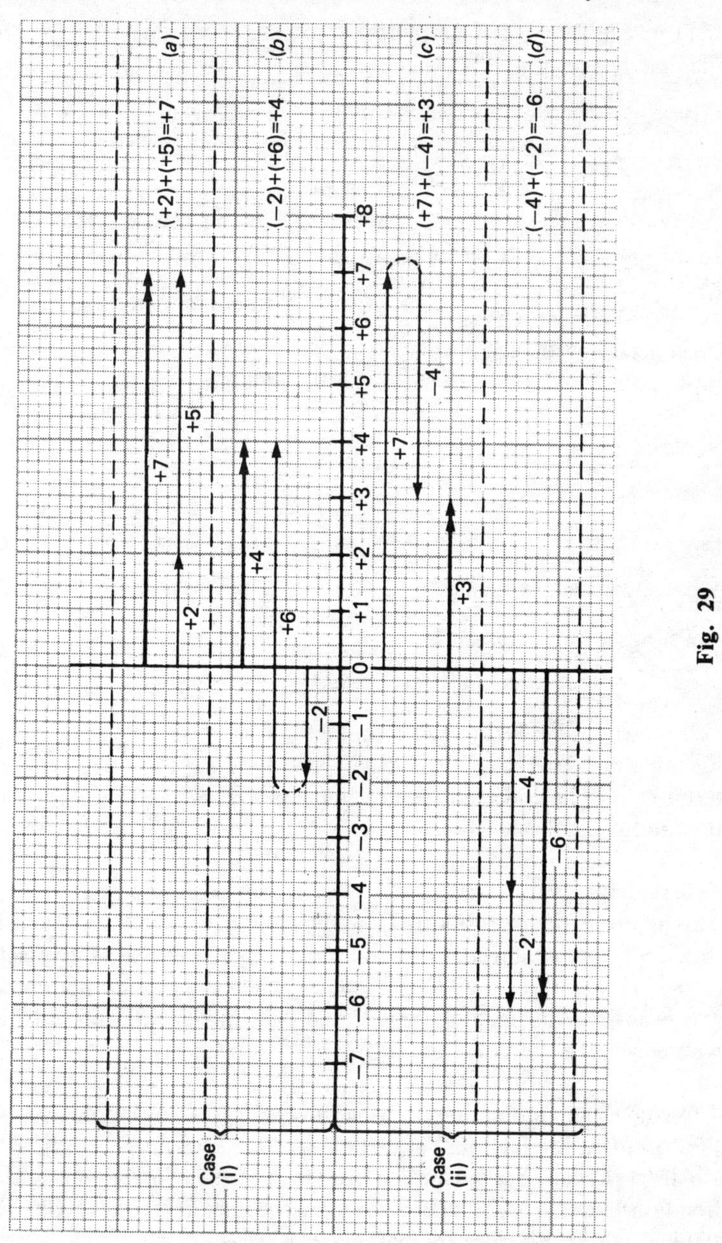

Fig. 29

From Fig. 29, we deduce an important pair of results with regard to the *doubly-signed* numbers.

Case (*i*), using the rules on pages 70–3 for singly-signed numbers:

(a) reduces to $2+(+5)=7$
Subtract 2 from each side
$\therefore \quad -\not{2}+\not{2}+(+5)=7-2$
i.e. $+(+5)=5$

(b) reduces to $-2+(+6)=4$
Add 2 to each side
$\therefore \quad \not{2}-\not{2}+(+6)=4+2$
i.e. $+(+6)=6$

We deduce that, in general, $+(+a)=a$, where a is any number.

Case (*ii*), using the same rules:

(c) reduces to $7+(-4)=3$
Subtract 7 from each side
$\therefore \quad -\not{7}+\not{7}+(-4)=3-7$
i.e. $+(-4)=-4$

(d) reduces to
$-4+(-2)=-6$
Add 4 to each side
$\therefore \quad \not{4}-\not{4}+(-2)=-6+4$
i.e. $+(-2)=-2$

We deduce that, in general, $+(-a)=-a$, where a is any number.

Taking a simple example:

$$3p+(+5p)+(-7p)=3p+5p-7p=p.$$

If, however, we had had different letters, e.g. $3p+(+5q)+(-7r)$, we could not have gone further than $3p+5q-7r$, as the 'p', 'q' and 'r' terms cannot be collected together (unless we have further information about their values).

Subtraction

This is rather more subtle than addition and we shall approach it differently. Let us consider three men climbing a staircase which starts in the basement of a building, 12 stairs below the ground floor, and proceeds up to the first floor which is 14 stairs higher still, and which is where the men are to assemble.

We take the number line to be *vertical*, the ground floor level as 0, the basement level as -12 and the first floor level as $+14$ (Fig. 30).

Albert (A) has already climbed from the basement, level -12, to the first floor, level $+14$; Bill (B) is taking a breather at stair $+4$; Charlie (C) is just starting from the basement, level -12.

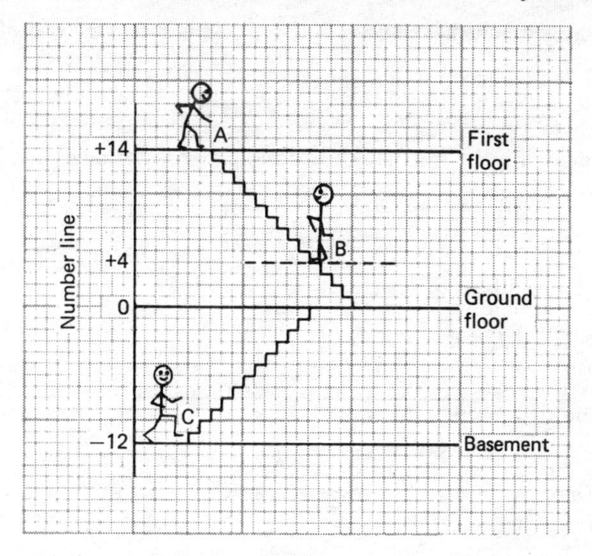

Fig. 30

Bill still has to climb $+14-(+4)$ stairs, which is clearly $14-4$ stairs, from the diagram; but $+14 = 14$,

hence $$14-(+4) = 14-4,$$

and on subtracting 14 from each side, we get

$$-(+4) = -4.$$

This method would have worked for any other number instead of -4,

$\therefore$ we deduce that, in general,

$$-(+a) = -a, \text{ where } a \text{ is any number.}$$

Charlie has to climb from -12 to $+14$, a change in level of $14-(-12)$, but his total ascent is $14+12$ stairs. From the diagram, therefore:

$$14-(-12) = 14+12, \text{ which gives } -(-12) = +12,$$

on subtracting 14 from each side.

This also would have succeeded for any number instead of -12,

$\therefore$ we deduce that, in general,

$$-(-a) = +a = a.$$

Summarising the four rules of double signs, we have:

$$\text{(i)}\quad +(+) = +, \qquad \text{(ii)}\quad +(-) = -,$$
$$\text{(iii)}\quad -(+) = -, \qquad \text{(iv)}\quad -(-) = +,$$

and when numbers or letters are attached:

$$\text{(i)}\quad +(+a) = +a = a; \qquad \text{(ii)}\quad +(-a) = -a;$$
$$\text{(iii)}\quad -(+a) = -a; \qquad \text{(iv)}\quad -(-a) = +a = a.$$

Taking two simple illustrations:

(a) $5pq-(-3pq) = 5pq+3pq = 8pq$;
(b) $-8m-(+7n) = -8m-7n$,; we cannot go any further, as m and n are different letters.

Example 1 (a) How much greater than -5 is 4?
(b) By how much does -3 exceed -10?
(c) By what amount is (i) 6°C, (ii) -6°C, less than 14°C?
(d) Find by how much a temperature of -12°C is higher than one of -8°C.

(a) The amount by which 4 is greater than -5 is
$4-(-5) = 4+5 = \mathbf{9}$.
(b) The amount by which -3 exceeds -10 is
$-3-(-10) = -3+10 = \mathbf{7}$.
(c) The amount by which 14°C is greater than
(i) 6°C is, in °C, $14-6 = 8$; i.e. the amount is **8°C**:
(ii) -6°C is, in °C, $14-(-6) = 20$; i.e. the amount is **20°C**.
(d) This is a trick question, because -12°C is a lower temperature than -8°C. There is, nevertheless, a sensible answer.
The number of degrees by which -12°C exceeds -8°C is, in °C,

$$-12-(-8) = -12+8 = -4$$

i.e. -12°C is higher than -8°C by **-4°C** (which means, of course, that the temperature, -12°C, is 4°C lower than -8°C).

Example 2 Solve the equation $2x-3(x-4)=8$.

We have $$2x-3x+12=8$$
$\therefore$ $$-x+12=8$$
giving $$-x=8-12=-4$$

Changing the signs of *both sides* of the equation does not affect its balance,

$\therefore$ $$-(-x)=-(-4)$$
i.e. $$\mathbf{x=4}$$

Example 3 Brown put $\frac{2}{5}$ of his invested capital into a building society. For financial reasons, he later decides to reduce the amount in the building society to $\frac{1}{4}$ of the original capital, and thereby finds that £600 ready cash will be released. How large was his original *total invested capital* and how much will he have left in the building society after making his withdrawal?

Let the original capital be £x, then the original amount in the building society was £$\frac{2}{5}x$, which has been reduced to £$\frac{1}{4}x$; the difference between the two is £600,

$\therefore$ $$\frac{2}{5}x-\frac{1}{4}x=600$$

Multiply through by the LCM of the denominator, i.e. 20.

$\therefore$ $$8x-5x=12000$$
and so $$3x=12000$$
$\therefore$ $$x=4000$$

The original capital was **£4000**; the amount *now* in the building society is £$\frac{1}{4}\times 4000=$ **£1000**.

EXERCISE 1

1 Simplify the following:

(a) $(-4)+(+7)$ (b) $(-3)-(+8)$
(c) $(+5)-(-6)$ (d) $(-12)-(-11)$
(e) $-(4)-(-4)$ (f) $(-2)+(-3)-(6-1)$
(g) $0-(8-12)$ (h) $3k+(-2k)$
(i) $4y-(y-7y)+(-3z)$ (j) $-(+3a)+(-2b)-(c-4c)$

2 A speculative gold mining share was erratically quoted during one day on the stock exchange. It opened at 75 points, rose by 32, only

to fall by 47; it then recovered by 24, from which it lastly dropped 16 points. Draw the right hand (positive) half of the number line, starting with 0 at the left hand end. (No stock is quoted at less than zero!). On the line mark the quotations at the correct positions, labelling them A, B, C. . . , in order of occurrence. Compare the final quotation with that obtained by finding the whole day's progress, as a single expression using plus and minus signs, and calculating the result. How many points is this above or below the initial quotation?

3 By how much does
(a) 21 °C exceed −7 °C, (b) −5 °C exceed −12 °C,
(c) −4 °C exceed 14 °C, (d) a distance of 17 km exceed 11 miles (take 1 km $\simeq \frac{5}{8}$ miles); answer in kilometres.

4 Solve the following equations:

(a) $2x-7=3$ (b) $6+5y=21$ (c) $2y+5=3y$
(d) $8+z=-2$ (e) $2-3y=7$ (f) $-5-z=-1$
(g) $\frac{1}{3}x-\frac{1}{4}=\frac{1}{6}$ (h) $3-\frac{2}{5}y=\frac{1}{10}$
(i) $2-\frac{4}{7}x=\frac{1}{2}x-1$ (j) $5(x-2)-2(x+4)=0$
(k) $t-3(1-3t)=10$ (l) $3(y-2)-(3-4y)=2(4y+1)$
(m) $\frac{1}{2}(q-1)=\frac{1}{3}(2q-\frac{3}{2})$ (n) $2(x+3)+3(4-\frac{2}{3}x)=0.$

Question (n) is included as a warning that not all straightforward-looking equations have a simple solution. Is there, in fact, a sensible answer for the value of x?

5 Because of the increased cost of petrol, a long distance driver reduces his average speed from 50 m.p.h. to 44 m.p.h. He finds that a particular journey takes 45 minutes longer than before. How long did it originally take him and how long is the journey? (*Hints*: Let the original time be t hours; distance travelled = speed (m.p.h.) × time (hours).)

4 Multiplication and division of positive and negative numbers

Multiplication

There are four cases: $(+a)(+b)$, $(+a)(-b)$, $(-a)(+b)$ and $(-a)(-b)$.

Let us go back for a moment to the four rules of double signs, given in the previous section. The rules are:

$$+(+a) = +a = a;\ +(-a) = -a;$$
$$-(+a) = -a;\ -(-a) = +a = a.$$

(i) The first case of multiplication, namely $(+a)(+b)$, is simple, for $+a = a$ and $+b = b$,

$$\therefore \qquad (+a)(+b) = ab.$$

For (ii), (iii) and (iv), however, we need one other rule as well. It is the Commutative Law of Multiplication (Ch. 3, Section 8, Law 2), which states briefly that $ab = ba$, where a and b are any numbers, positive, negative or zero. Hence we get

(ii) $(+a)(-b) = (-b)(+a) = (-b)a$ (law 2 above, also $+a = a$)
$\qquad = -ba$ (rules of double signs)
$\qquad = -ab$ (law 2)

(iii) $(-a)(+b) = (-a)b = -ab$ (as $+b = b$)

(iv) $(-a)(-b) = -a(-b) = -(-b)a$ (law 2)
$\qquad = +ba$ (rules of double signs)
$\qquad = ab$, the same result as in (i)

Summarising,
$$(+a)(+b) = (-a)(-b) = ab$$
$$(+a)(-b) = (-a)(+b) = -ab.$$

These important results for multiplication are really an extension of the rules of double signs.

Special cases: If we put $b = a$ in the summary above, and remember (page 7) that $a \times a = a^2$, then

$$(+a)(+a) = (-a)(-a) = a^2$$
$$(+a)(-a) = (-a)(+a) = -a^2$$

Furthermore, if $b = a = 1$, this reduces to

$$(+1)(+1) = (-1)(-1) = 1^2 = 1$$
$$(+1)(-1) = (-1)(+1) = -(1)^2 = -1$$

This can be extended indefinitely:

$$(+1)(+1)(+1) = \{(+1)(+1)\}(+1) = (+1)(+1) = 1$$
$$(-1)(-1)(-1) = \{(-1)(-1)\}(-1) = (+1)(-1) = -1$$

i.e. $(+1)^3 = +1$; $(-1)^3 = -1$
$(+1)^4 = +1$; $(-1)^4 = +1$
$(+1)^5 = +1$; $(-1)^5 = -1$, and so on.

The rule follows that

(i) If all the numbers multiplied together are positive (i.e. they all have plus signs), the result is *positive*; e.g. $(+k)^3 = k^3$, $(+p)^4(+q)^3 = p^4q^3$.
(ii) If some of the numbers multiplied together are positive (+ signed) and the rest are negative (− signed), the result is

(a) *positive*, if the number of negative signs is *even*,
e.g. $(+y)^4(-z)^6 = y^4z^6$;
(b) *negative*, if the number of negative signs is *odd*,
e.g. $2^2(-d)^3 = -4d^3$.

Further examples:

$$(+2)^3(+3)^2 = 2^3 \times 3^2 = 2 \times 2 \times 2 \times 3 \times 3$$
$$= 8 \times 9 = 72 \qquad \text{(all + signs)}$$

$$(-3)^4 = 3 \times 3 \times 3 \times 3 = 81 \qquad \text{(4 − signs)}$$

$$(-5)^3x^2 = -125x^2 \qquad \text{(3 − signs)}$$

$$5(-2p)(-3q)(+4r)^2 = 5 \times 2p \times 3q \times 4r \times 4r$$
$$= 480pqr^2 \qquad \text{(2 − signs in all)}$$

$$\left(-\frac{1}{2}\right)^3\left(-\frac{2}{3}x\right) = +\frac{1}{\not{8}_4} \times \frac{\not{2}^1x}{3} = \frac{x}{12} \qquad \text{(4 − signs in all)}$$

Division

As with multiplication, there are four cases:

$$(+a) \div (+b),\ (+a) \div (-b),\ (-a) \div (+b),\ (-a) \div (-b)$$

All four cases can be directly derived from earlier ideas. We must firstly remember that if we multiply the numerator *and* the denominator of a fraction by the same number, letter or combination of these, the fraction is unaltered.

Hence

(i) $$(+a) \div (+b) = \frac{+a}{+b} = \frac{a}{b}$$

(ii) $$(+a) \div (-b) = \frac{+a}{-b} = \frac{(+a)(-1)}{(-b)(-1)}$$

$$= \frac{-a}{+b} = -\frac{a}{b}$$

(using *multiplication* rules above)

(iii) $$(-a) \div (+b) = \frac{-a}{+b} = -\frac{a}{b}$$

(iv) $$(-a) \div (-b) = \frac{-a}{-b} = \frac{(-a)(-1)}{(-b)(-1)}$$

(again using multiplication rules)

$$= \frac{+a}{+b} = \frac{a}{b}$$

It follows that the *rules for signs in division are the same as the rules for multiplication* in all the cases (i) to (iv), and that extensions can also be developed in a similar way.

Example 4 Simplify (a) $(-18) \div (-3)$, (b) $(-24) \div (+16)$, (c) $8ab \div (-6db)$, (d) $(-3p) \times (-6q) \div (4pr)$

We have

(a) $(-18) \div (-3) = 18 \div 3 = \mathbf{6}$

(b) $(-24) \div (+16) = -\frac{24}{16} = -\frac{\mathbf{3}}{\mathbf{2}}$ (on cancelling)

(c) $8ab \div (-6db) = -\frac{8ab}{6db} = -\frac{\mathbf{4a}}{\mathbf{3d}}$

(d) $(-3p) \times (-6q) \div 4pr = \frac{18pq}{4pr} = \frac{\mathbf{9q}}{\mathbf{2r}}$

Example 5 If $a = -2$ and $b = -\frac{3}{4}$, find the values of (a) $7a \div 6b$, (b) $-5ab \div (-\frac{2}{3}b)$

(a) We have

$$\frac{7a}{6b}=\frac{7\times(-2)}{6\times\left(-\frac{3}{4}\right)}=\frac{7\times 2}{6\times\frac{3}{4}}=\frac{7\times 2\times\frac{4}{3}}{6\times\frac{3}{4}\times\frac{4}{3}}=\frac{7\times 2}{6}\times\frac{4}{3}\left(\text{as }\frac{3}{4}\times\frac{4}{3}=1\right)$$

$$=\frac{28}{9}=\mathbf{3\frac{1}{9}}.$$

(b) Also

$$\frac{-5ab}{-\frac{2}{3}ab}=\frac{5a}{\frac{2}{3}}\text{ (on cancelling the } b\text{s)}$$

$$=\frac{5(-2)\times\frac{3}{2}}{\frac{2}{3}\times\frac{3}{2}}=-10\times\frac{3}{2}=\mathbf{-15.}$$

EXERCISE 2

1 Simplify

(a) $(+6)\times(-5)$ (b) $(-6)^2\div(+3)$ (c) $(-48)\div(-8)^3$

(d) $(-9)\div(+63)$ (e) $(2a)\times(-6a)$

(f) $(-3bc)\times(-4d)$ (g) $(-12ab)\div(+16bc)$

2 Find the value of the following, when $x=-3$, $y=4$,

(a) $5x\div 9y$ (b) $(+2x)\times(-3y)$ (c) $\left(-\frac{1}{2}x\right)\times(5y)\div(6xy)$

5 Indices extended to algebra

Before we can apply the foregoing basic processes of algebra to more elaborate uses of brackets, we need to extend the theory of indices in two ways, firstly with regard to letters (generalised numbers), and secondly for negative, as well as positive, indices.

The first extension is easy, for the rules are virtually unchanged, for example:

$$7^4=7\times 7\times 7\times 7$$

and similarly

$$a^4=a\times a\times a\times a$$

(as explained on page 20)

Addition of indices

The rules follow the same pattern as in arithmetic (cf. page 20).

Examples:

$$a^5 \times a^3 = a^{5+3} = a^8 \qquad \text{(as for numbers)}$$
$$3b^2 \times b^4 = 3 \times b^2 \times b^4 = 3b^{2+4} = 3b^6 \qquad \text{(the 3 is unaffected)}$$
$$7c \times 5c^3 = 7 \times 5 \times c \times c^3 = 35c^{1+3} = 35c^4$$
$$\text{(or } 7.5.c.c^3 = 35c^4\text{, which is shorter)}$$
$$4a^2b \times 6b^3c = 4.6a^2b.b^3c = 24a^2b^4c$$
$$2a^4b \times 3abc \times \frac{5}{6}b^2d^2 = 2 \times 3 \times \frac{5}{6} \times a^{4+1} \times b^{1+1+2} \times c \times d^2$$
$$= 5a^5b^4cd^2$$

To summarise, if m and n are positive integers (whole numbers) and a is any real number,

$$\boldsymbol{a^m \times a^n = a^{m+n}}$$

Multiplication of indices

Again the process is much the same as in arithmetic, but this time multiplication has been intentionally placed before subtraction as we shall be studying the *latter* in greater detail. When brackets occur, any power to which the bracket is raised affects the whole of the contents thereof, e.g. $(p^4)^5 = p^{4\times 5} = p^{20}$; $(4a^2)^3 = 4^3a^{2\times 3} = 64a^6$.

Examples:

$$(xy)^3 = (xy) \times (xy) \times (xy) = (x \times x \times x) \times (y \times y \times y) = x^3y^3$$
$$(2pq^3)^2 = 2^2 \times p^2 \times (q^3)^2 = 4p^2q^{3\times 2} = 4p^2q^6$$
$$6\left\{\frac{2}{3}h\right\}^3 = 6 \times \left\{\frac{2}{3}\right\}^3 \times h^3 = 6 \times \frac{8}{27} \times h^3 = \frac{16}{9}h^3.$$

(Note that the number 6 outside the bracket is unaffected by the power of the bracket.)

To summarise, if m and n are positive integers, and a is any real number,

$$\boldsymbol{(a^m)^n = a^{m\times n} = a^{mn}}$$

Subtraction of indices and extension to negative indices

Let us now investigate what happens when we consider negative indices. We firstly take the case of subtraction, instead of addition. The use originally arose on page 20, when we investigated the values of $6^5 \div 6^3 = 6^2$ and $6^{5-3} = 6^2$ and we deduced that $6^5 \div 6^3 = 6^{5-3}$. The statement then made was that this result is generally true, i.e. if a power m of a number is divided by another power, n, of the same number, the result is the original number raised to the power $m-n$, provided that m and n are integers, hence

$$8^m \div 8^n = 8^{m-n}$$

We can immediately extend this for letters; for example:

$$a^7 \div a^4 = \frac{a \times a \times a \times a \times a \times a \times a}{a \times a \times a \times a} = a \times a \times a = a^3$$

i.e. $\quad a^7 \div a^4 = a^{7-4} = a^3$ is true

Similarly, for any integers m and n, if m is greater than n (which is written $m > n$), we have $a^m \div a^n = a^{m-n}$

What happens, however, if m is less than n (which is written $m < n$)? This would mean $m-n$ is negative. Look at the following:

$$a^2 \div a^5 = \frac{a \times a}{a \times a \times a \times a \times a} = \frac{1}{a \times a \times a} = \boxed{\frac{1}{a^3}}$$

but if our rule of subtraction is to hold, we need to have

$$a^2 \div a^5 = a^{2-5} = \boxed{a^{-3}}$$

Comparing the results inside the dotted rectangle, we conclude that we can use the negative index, -3, provided that we *define* a^{-3} as meaning $1 \div a^3$. This result is also true in the general case provided that m and n are positive integers and a is any real number (but beware of the case $a = 0$, which presents special difficulty).

$$\boldsymbol{a^m \div a^n = a^{m-n} \text{ and } a^{-n} = \frac{1}{a^n}}$$

There is, however, a particular case which has not yet received mention, namely the odd situation which occurs if $m = n$. This would lead to

$$a^m \div a^m = a^{m-m} = a^0$$

using the above procedure. We need to find out whether this has a meaning.

Consider, say,

$$a^3 \div a^3 = \frac{a \times a \times a}{a \times a \times a} = 1, \text{ as all the } as \text{ cancel out.}$$

Likewise, in general,

$$a^m \div a^m = \frac{a \times a \times a \times \ldots a \ (m \text{ times})}{a \times a \times a \times \ldots a \ (m \text{ times})} = 1 \text{ as again, all } as \text{ cancel out.}$$

$\therefore$ we have $a^m \div a^m = a^0$ and $a^m \div a^m = 1$

Hence our formula holds in this special case, if and only if we *define*

$$a^0 = 1.$$

Examples:

$$p^{10} \div p^7 = p^{10-7} = p^3$$

$$x^4 \div x^9 = x^{4-9} = x^{-5} = \frac{1}{x^5}$$

$$a^6 \div (a^2)^3 = a^6 \div a^6 = a^{6-6} = a^0 = 1$$

$$(2q)^{-3} = \frac{1}{(2q)^3} = \frac{1}{8q^3}$$

$$\frac{5}{2}y \times (6y^2)^2 \div (3y)^3 = \frac{5}{2}y \times 36y^4 \times \frac{1}{27y^3} = \frac{10}{3}y^{1+4-3} = \frac{10}{3}y^2$$

Division of indices

The fourth (and last) case is $(a^m)^{\frac{1}{n}} = a^{\frac{m}{n}}$; *we shall only consider simple examples.*

Definition The *square root* of a number is another number which when multiplied by itself gives the original number. The symbol for a square root is $\sqrt{\ }$, e.g. $\sqrt{3}$; $\sqrt{428}$.

Some numbers, known as *perfect squares*, have exact square roots. The simplest ones are easily found, because $1 \times 1 = 1$, $2 \times 2 = 4$,

$3 \times 3 = 9$, $4 \times 4 = 16$, $5 \times 5 = 25$, etc. The sequence of numbers 1, 4, 9, 16, 25 . . . consists of perfect squares, and clearly $\sqrt{1} = 1$, $\sqrt{4} = 2$, $\sqrt{9} = 3$, $\sqrt{16} = 4$, $\sqrt{25} = 5$, etc., these being exact square roots. Most numbers do not, however, have exact square roots, for example:

$$\sqrt{2} \times \sqrt{2} = 2, \text{ but } \sqrt{2} \text{ is not exact: } \sqrt{2} \simeq 1.4142$$

likewise $\sqrt{3} \times \sqrt{3} = 3$, where $\sqrt{3} \simeq 1.7321$

Let us now consider what is meant by a simple case of $(a^m)^{\frac{1}{n}}$. We put $a = 3$, $m = 1$, $n = 2$ and the expression reduces to $3^{\frac{1}{2}}$ (i.e. 3 to the power $\frac{1}{2}$). Let us square it and assume that the rules for indices still apply, then

$$3^{\frac{1}{2}} \times 3^{\frac{1}{2}} = 3^{\frac{1}{2}+\frac{1}{2}} = 3^1 = 3 \quad . \quad . \quad . \quad . \quad . \quad (1)$$

but by definition, $\sqrt{3} \times \sqrt{3} = 3 \quad . \quad . \quad . \quad . \quad . \quad (2)$

$\therefore$ by comparison between (1) and (2)

$$3^{\frac{1}{2}} = \sqrt{3}$$

and more generally $a^{\frac{1}{2}} = \sqrt{a}$, where a is a *positive* number.

Definition The nth root of a number is another number which, when put down n times, these being multiplied together give the original number.

For example:

$$\sqrt[3]{a} \times \sqrt[3]{a} \times \sqrt[3]{a} = a \qquad \text{(definition)}$$

but $a^{\frac{1}{3}} \times a^{\frac{1}{3}} \times a^{\frac{1}{3}} = a^{\frac{1}{3}+\frac{1}{3}+\frac{1}{3}} = a^1 = a$

(as rule of indices applies)

$$\therefore \quad a^{\frac{1}{3}} = \sqrt[3]{a}$$

(same method as above)

We deduce that in the general case, the meaning of $(a^m)^{\frac{1}{n}}$ is given by

$$(a^m)^{\frac{1}{n}} = a^{\frac{m}{n}} = \sqrt[n]{a^m}$$

i.e. the nth root of a to the power m, where we shall take m and n as positive integers and a as a positive number. (Those who wish to take a as negative are referred to *New Mathematics* (Teach Yourself Books), Chapter 2; there are complications involving $\sqrt{-1}$.)

Now this is all very jolly, but we are not yet quite out of the wood. We must consider our rules of signs with regard to square roots.

(Other *even-indexed* roots, i.e. 4th, 6th, etc., are likewise affected.) We know that $(+)(+) = +$ and $(-)(-) = (+)$, hence if we take, say, the equation $x^2 = 3$, we have $x = 3^{\frac{1}{2}} = \sqrt{3}$, but $(+\sqrt{3})(+\sqrt{3}) = 3$ and $(-\sqrt{3})(-\sqrt{3}) = 3$, therefore $x = +\sqrt{3}$ or $-\sqrt{3}$;

$\therefore$ 3 has two *different* square roots, $+\sqrt{3}$ and $-\sqrt{3}$.

This applies to all square roots, whether we are dealing with numbers or letters.

In *arithmetic*, we frequently ignore the negative root, as we are mainly concerned with positive answers, but in *algebra* we must consider carefully whether or not we need the negative result as well as, or instead of, the positive result. Hence, using the symbol $\pm$ to mean 'plus or minus', we have:

$$9^{\frac{1}{2}} = \pm 3,\ 5^{\frac{1}{2}} = \pm\sqrt{5} \simeq \pm 2.236,\ a^{\frac{1}{2}} = \pm\sqrt{a},\ (4p^6)^{\frac{1}{2}} = \pm 2p^3,$$

and slightly harder:

$$(49p^3)^{\frac{1}{2}} = +7p^{\frac{3}{2}} = \pm 7\sqrt{p^3}.$$

Consolation may be drawn from the fact that *odd*-indexed roots have only one *real* value, for example:

$$8^{\frac{1}{3}} = 2,\ 7^{\frac{1}{3}} = \sqrt[3]{7},\ (27h^6k^3)^{\frac{1}{3}} = 3h^2k.$$

Example 6 Find the second and third powers of (a) $-\frac{2}{3}$, (b) $3a^2$, (c) $-\sqrt{6}$

We have

(a) $\left(-\frac{2}{3}\right)^2 = +\left(\frac{2}{3}\right)^2 = \mathbf{\frac{4}{9}}$ (2 minus signs);

$\left(-\frac{2}{3}\right)^3 = -\left(\frac{2}{3}\right)^3 = -\mathbf{\frac{8}{27}}$ (3 minus signs)

(b) $(3a^2)^2 = 3^2(a^2)^2 = \mathbf{9a^4}$; $(3a^2)^3 = 3^3(a^2)^3 = \mathbf{27a^6}$

(c) $(-\sqrt{6})^2 = +(6^{\frac{1}{2}})^2 = 6^{\frac{1}{2}\times 2} = 6^1 = \mathbf{6}$;

$(-\sqrt{6})^3 = -(6^{\frac{1}{2}})^3 = -6^{\frac{3}{2}} = -6.6^{\frac{1}{2}} = \mathbf{-6\sqrt{6}}$.

(*Note*: $6\sqrt{6}$ means $6 \times \sqrt{6}$, *not* the sixth root of 6 which is written $\sqrt[6]{6}$.)

Example 7 Simplify (a) $7a^2b \times 3b^3$, (b) $(3k^4)^2$, (c) $6p^{-2}$, (d) $(5t^2)^{-1}$, (e) the product of $4m^2n$ and $5mn^{-3}$, (f) $c^2d^3 \times cd^{-3}$, (g) the cube of $27y^2$.

We have

(a) $7a^2b \times 3b^3 = 7 \times 3a^2b \times b^3 = \mathbf{21a^2b^4}$; (b) $(3k^4)^2 = 9k^{4\times 2} = \mathbf{9k^8}$

(c) 'product of' means 'multiplication of', hence

$$4m^2n \times 5mn^{-3} = 4 \times 5m^2n \times mn^{-3} = 20m^3n^{1-3} = \frac{\mathbf{20m^3}}{\mathbf{n^2}}.$$

(d) $6p^{-2} = \dfrac{\mathbf{6}}{\mathbf{p^2}}$ (the 6 is unaffected here)

(e) $(5t^2)^{-1} = \dfrac{\mathbf{1}}{\mathbf{5t^2}}$ (as the 5 is inside the bracket with the t^2)

(f) $c^2d^3 \times cd^{-3} = c^2 \times c \times d^3 \times d^{-3} = c^3d^0 = \mathbf{c^3}$ (as $d^0 = 1$)

(g) The cube root of $27y^2$ is

$$(27y^2)^{\frac{1}{3}} = (3^3)^{\frac{1}{3}}(y^2)^{\frac{1}{3}} = 3^{3\times\frac{1}{3}}y^{\frac{2}{3}} = 3y^{\frac{2}{3}} = \mathbf{3\sqrt[3]{y^2}}.$$

Example 8 Find the values of $(b^2 - 4ac)^{\frac{1}{2}}$, when $a = 6, b = -7$ and $c = 1$.

We have

$$b^2 - 4ac = (-7)^2 - 4 \times 6 \times 1 = 49 - 24 = 25$$

$$\therefore \quad (b^2 - 4ac)^{\frac{1}{2}} = \pm 25^{\frac{1}{2}} = \mathbf{\pm 5}$$

This is clearly correct because $5 \times 5 = 25$ and $(-5)(-5) = 25$, but the question brings out an important point, namely, if a is any positive number,

$$a^{\frac{1}{2}} = \pm\sqrt{a}, \textit{ but } \sqrt{a} \text{ means } +\sqrt{a} \textit{ only};$$
$$\text{likewise, } -\sqrt{a} \text{ means } -\sqrt{a} \text{ only.}$$

EXERCISE 3

1 Simplify: (a) $2ab - 5a + 3b - 6ab$, (b) $7abc - 5cab - (-acb)$, (c) $(2xy - yz + zx) - (2xy + 3xz - 4xy) + 3y(z - x)$. (Remember the commutative and associative laws given in Chapter 3).

2 Reduce to simplest form:

(a) $2y^2 \times y^3$, (b) $(2x)^2 \times y^3$, (c) $5n^2 \times \frac{3}{n^3}$, (d) $3a^2b \div 12ab^2$,

(e) $\frac{3}{7}x^4 \times \left(\frac{2x}{3}\right)^3$, (f) $(-2p^2)\left(-\frac{7}{p}\right)$, (g) $3s(-2t)-4t(-5s)$,

(h) $3y^2-6y(\frac{1}{2}y-\frac{1}{3}y)$, (i) $(3x)^{-1}$, (j) $3x^{-1}$, (k) $9(6x)^{-2}$,

(l) $16^{\frac{1}{4}}$, (m) $4^{\frac{3}{2}}$, (n) $\left(\frac{27}{8}\right)^{\frac{1}{3}}$, (o) $\left(\frac{2}{5}\right)^{-1}$, (p) $\left(\frac{c}{2d}\right)^{-2}$, (q) $(z^{-2})^{-3}$.

3 Write down the third and fourth powers of (a) -3, (b) m^2, (c) $2a^{\frac{1}{2}}$, (d) $\sqrt{7}$, (e) $-\frac{t}{\sqrt{3}}$.

4 Put down the first, second, third, fourth and fifth powers of $\frac{1}{2}$, and add them together. Now add the next power of $\frac{1}{2}$. What do you notice?

If we continue to add higher and higher powers of $\frac{1}{2}$, what do you think will be the *ultimate* sum of $\frac{1}{2}+\frac{1}{2^2}+\frac{1}{2^3}+\frac{1}{2^4}+\dots$, assuming that we go on for ever?

5 Simplify:

(a) $\frac{3x}{4}-\frac{5x}{6}$, (b) $\frac{2s}{5}-\left(\frac{3s}{4}-\frac{2s}{3}\right)$.

(*Hint*: These follow the same rules for fractions as in Chapter 2, section 10).

6 Find the values of the following:

(a) a^2+b^2, when $a=2$, $b=-3$,
(b) $f^3+g^3-h^3$, when $f=1$, $g=-2$, $h=-3$,

(c) $(4d)^{-1}$ when $d=\frac{3}{2}$, (d) $3r^2-5r+7$, when $r=-\frac{1}{3}$,

(e) $(5p^2q)^2$, when $p=\frac{1}{2}$, $q=\frac{2}{5}$,

(f) $\sqrt{-12h^2k}$, when $h=-\frac{1}{2}$, $k=-3$, (g) $4u^{-3}$, when $u=6$.

7 The equation $ax^2+bx+c=0$ can be solved. It gives two possible

values of x in terms of a, b and c. These values are:

$$x = \frac{-b+\sqrt{b^2-4ac}}{2a} \text{ or } \frac{-b-\sqrt{b^2-4ac}}{2a}$$

If $a = 2$, $b = 3$ and $c = -2$, find the possible values of x (Look back at Example 8 just above Exercise 3). (The equation $ax^2+bx+c=0$ is known as a *quadratic equation*, i.e. one of the second degree, in x; ax^2 is of the second degree in x, bx is of the first degree and c is the independent term.)

Partial solution, left for the reader to complete:

$$a = 2, b = 3, c = -2$$

$$\therefore \quad x = \frac{-3+\sqrt{3^2-4.2.(-2)}}{2.2} = \frac{-3+\sqrt{9+16}}{4}$$

$$= \frac{-3+\sqrt{25}}{4} = \frac{-3+5}{4}, \text{ etc.}$$

$$\text{or} \quad x = \frac{-3-\sqrt{3^2-4.2(-2)}}{2.2} = \text{etc.}$$

8 Using the method of question 7 above, solve the equations (a) $2x^2+7x+5=0$, (b) $2x^2-x-2=0$, given that $\sqrt{17} \simeq 4.123$.

6 Elementary – or is it?

The thoughtful reader may have considered it peculiar that, for any positive value of a (i.e. $a > 0$), the expression $a^{\frac{1}{2}}$ gives two *real* values, namely $\sqrt{a}$ and $-\sqrt{a}$, whereas $a^{\frac{1}{3}}$ yields only one real result, which is $\sqrt[3]{a}$. As $a^{\frac{1}{2}}$ yields *two* results, may not $a^{\frac{1}{3}}$ present us with three? This is indeed the case, but two of the values of $a^{\frac{1}{3}}$ are non-real.

Consider two simple equations, (i) $x^2 = 2$, (ii) $x^3 = 8$.

(i) gives (in full) $x^2 = 2 \Rightarrow (x^2)^{\frac{1}{2}} = 2^{\frac{1}{2}} \Rightarrow x^{2\times\frac{1}{2}} = 2^{\frac{1}{2}} \Rightarrow x = 2^{\frac{1}{2}}$

i.e. $\quad x = \sqrt{2}$ or $x = -\sqrt{2}$ (2 *real* roots)

(ii) gives $\quad x^3 = 8 \Rightarrow (x^3)^{\frac{1}{3}} = 8^{\frac{1}{3}} \Rightarrow x = 8^{\frac{1}{3}}$

i.e. $\quad x = \sqrt[3]{8} = 2$ (1 *real* root)

Actually (ii) *does* have two other roots, namely,

$$x = -\tfrac{1}{2} + \tfrac{1}{2}i\sqrt{3} \quad \text{or} \quad x = -\tfrac{1}{2} - \tfrac{1}{2}i\sqrt{3} \text{ (where } i = \sqrt{-1}\text{)}$$

but they are *complex* numbers, not *real* ones, and they are not suited to basic mathematical study. (See *New Mathematics*, Teach Yourself Books, Chapter 2.)

There is, in fact, a *fundamental theorem* which implies that every equation of the nth degree has exactly n roots. The general equation of the nth degree in x can be written

$$a_n x^n + a_{n-1} x^{n-1} + \ldots + a_1 x + a_0 = 0$$

where every $a_0, a_1 \ldots a_n$ is the coefficient of the corresponding power of x (the *suffixes* merely distinguishing between the different *as*). The values of x may then be written as:

$$x = \alpha_1, \alpha_2, \alpha_3 \ldots \alpha_n \qquad \text{(Greek } alpha\text{)}$$

where the alphas have to be determined; but, frequently, many of these roots are complex numbers. Roots may be repeated, e.g. those of $x^3 - 3x + 2 = 0$ are 1, 1 and -2.

The fundamental theorem of algebra, just mentioned, was so difficult to prove that for about 200 years mathematicians failed to find a successful solution. It was ultimately mastered by Carl Friedrick Gauss (1777 – 1855), an outstanding German mathematician, in 1799.

To enter into a discussion as to whether any number is *real*, considering that its concept is *abstract*, would lead us into the field of metaphysics – or at least into logical debate. Somebody called the *simple* numbers *real*, and the name has stuck! Let us leave it at that.

6

Triangles and Constructions

1 The triangle

The origin of the triangle is lost in the mists of antiquity. It is found as a form of decoration on ancient Sumerian pottery dating from 3500 BC, and contemporarily in Bohemia, which by then had reached its Bronze Age. The triangle was therefore well established as an art-form, and probably had astrological or mystical significance as well. At the same time it was, in conjunction with other geometrical figures, coming into use for such practical purposes as land measurement and building; by 2700 BC it was certainly the shape used for pyramidal faces. In mathematical history one frequently finds aesthetic and utilitarian approaches developing side by side.

Definition 20 Two geometrical figures are said to be *congruent* if they are alike in all respects. This implies that, although rarely done, one of them could be placed to fit exactly on top of the other (i.e. superposed), although it might have to be inverted (turned upside down) to do so. The property of congruence is of particular importance when associated with triangles.

The symbol $\equiv$ stands for 'is congruent with', and $\triangle$ represents the word 'triangle'.

Fig. 31 illustrates the two cases of superposition when two triangles are alike in all respects.

For those readers interested in logical thought, Fig. 31 presents a flaw with regard to classical plane geometry, in which everything is meant to be in a *plane* (i.e. a perfectly flat surface). It is necessary to lift, mentally, the $\triangle$ABC into space, in *Case I*, in order to lay it on $\triangle$XYZ; in *Case II* it is worse, as we also have to turn $\triangle$ABC upside down in

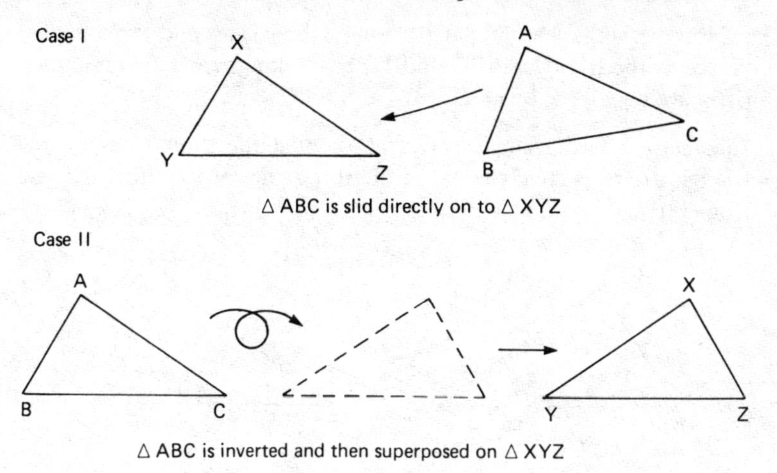

△ ABC is slid directly on to △ XYZ

△ ABC is inverted and then superposed on △ XYZ

Fig. 31

space! Fortunately, in real life, it is of no consequence.

The following theorems state four different sets of conditions, *each* of which will ensure that two triangles are congruent. *Proofs* of these four theorems are not given.

Theorem 3. Two triangles are congruent if two sides and the included angle (between them) of one triangle are equal, each to each, to two sides and the included angle of the other triangle (Fig. 32).

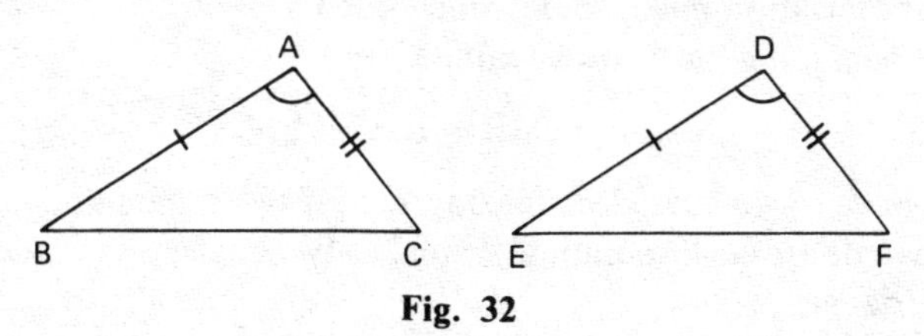

Fig. 32

The following is *not* a proof. It shows how the theorem is applied, where relevant in solving a problem.

If in △s ABC, DEF: AB = DE
AC = DF
$\hat{A} = \hat{D}$ (included angle) } given in Fig. 32
then △ABC ≡ △DEF(SAS)
(The abbreviation means Side, included Angle, Side.)

Now a triangle has six parts, namely, three sides and three angles. Hence as the triangles ABC and DEF are congruent, the remaining parts are equal, i.e. BC = EF, $\hat{B} = \hat{E}$ and $\hat{C} = \hat{F}$.

Theorem 4 Two triangles are congruent if the three sides of one triangle are respectively equal to the three sides of the other triangle (Fig. 33).

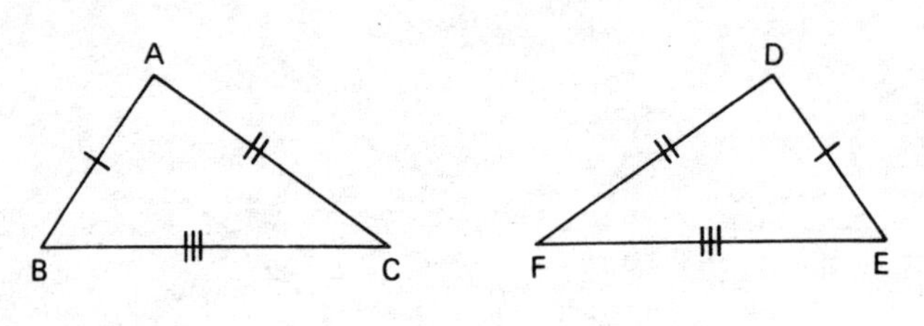

Fig. 33

When applying the theorem to a specific problem, the style of wording could be:

In △s ABC, DEF AB = DE, AC = DF and BC = EF } given in Fig. 33
∴ △ABC ≡ △DEF (SSS)

(The abbreviation means Side, Side, Side.)
∴ the remaining parts are equal, i.e.

$$\hat{A} = \hat{D},\ \hat{B} = \hat{E},\ \hat{C} = \hat{F}.$$

Theorem 5 Two triangles are congruent if two angles and a side of one triangle are correspondingly equal to two angles and a side of the other triangle.

The word 'correspondingly' is important here, for the equal sides must have the same positions relative to the equal angles (Fig. 34).

When using the theorem for a particular problem, it would be in the following style:

In △s ABC, DEF $\hat{A} = \hat{D}$ and $\hat{B} = \hat{E}$
BC = EF (corresponding) } given in Fig. 34
∴ △ABC ≡ △DEF (AA Cor. S)

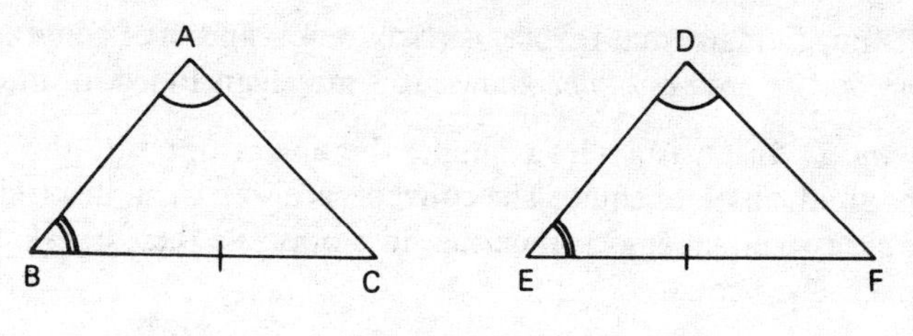

Fig. 34

(The abbreviation means Angle, Angle, Corresponding Side.)

$\therefore$ $AB = DE$, $AC = DF$ and $\hat{C} = \hat{F}$

Theorem 6 Two right-angled triangles are congruent if the hypotenuse and one other side of one triangle are respectively equal to the hypotenuse and one other side of the other triangle (Fig. 35). (N.B. This theorem applies only to *right-angled* triangles.)

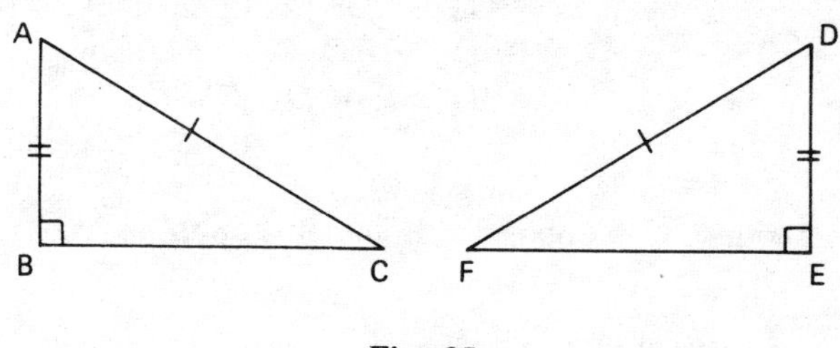

Fig. 35

The same comment applies as in the previous three theorems, when utilising this one. The following is merely a statement of how to use it.

In $\triangle$s ABC, DEF $\quad AB = DE$ (Side) $\quad AC = DF$ (Hypotenuse) $\Big\}$ given in Fig. 35

$\hat{B} = 90^\circ = \hat{E}$

$\therefore$ $\triangle ABC \equiv \triangle DEF$ (RHS)

(The abbreviation means Right angle, Hypotenuse, Side.)

$\therefore$ $BC = EF$, $\hat{A} = \hat{D}$ and $\hat{C} = \hat{F}$.

Definition 21 An isosceles triangle is one in which two sides are equal.

The following theorem *is* proved as an illustration of the use of one of the four fundamental theorems (Nos. 3, 4, 5 and 6) of congruence. *The proof does not need to be known*; it is merely included for interest.

Theorem 7 In an isosceles triangle, the angles opposite the equal angles are themselves equal. The converse is also true: if, in a triangle, there are two equal angles, then the sides opposite these equal angles are themselves equal.

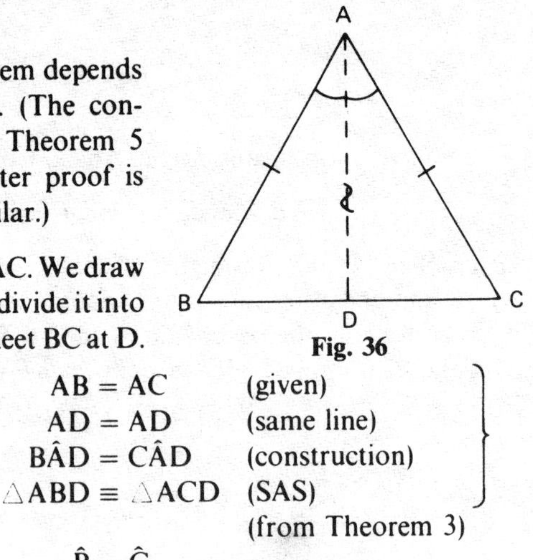

Fig. 36

The proof of the theorem depends upon Theorem 4 (SAS). (The converse entails the use of Theorem 5 (AA Cor S), but the latter proof is omitted as it is very similar.)

Proof In Fig. 36, AB = AC. We draw AD to bisect $C\hat{A}B$ (i.e. to divide it into two equal parts) and to meet BC at D.

In △sABD and ACD   AB = AC (given)
AD = AD (same line)
$B\hat{A}D = C\hat{A}D$ (construction)
∴ △ABD ≡ △ACD (SAS)
(from Theorem 3)

∴ $\hat{B} = \hat{C}$

where, of course, $\hat{C}$ is opposite AB and $\hat{B}$ is opposite AC.

EXERCISE 1

A. In each of the following pairs of triangles, write down whether they are or are not congruent.

B. In cases where they are congruent, give the lettering of the triangles in *corresponding* order. (In question 1, for example, △ABC ≡ △DFE, *not* △DEF.) Give the correct reason for congruence, e.g. (SAS), which does not happen to be correct in question 1!

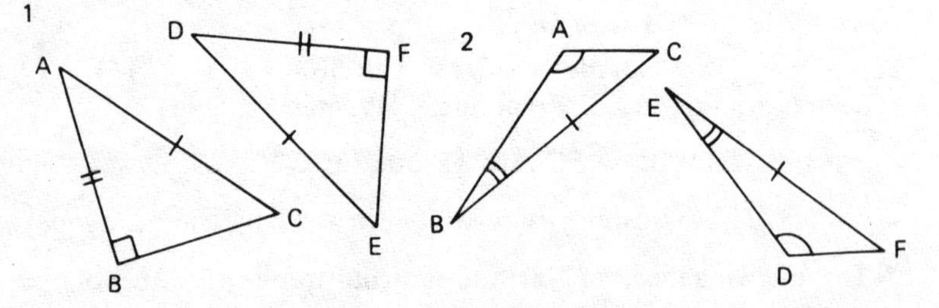

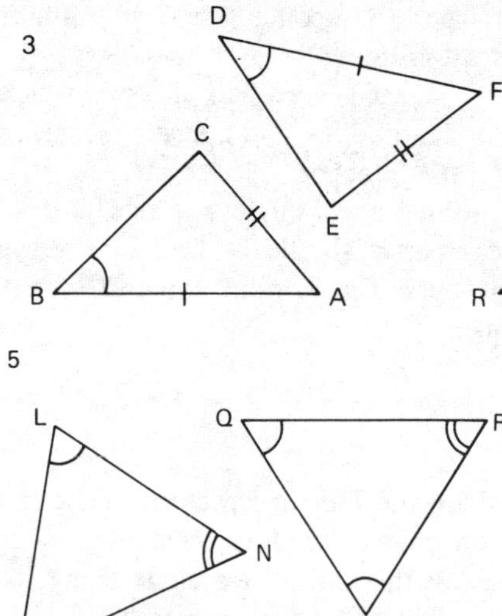

$\hat{L} = \hat{M} = \hat{P} = \hat{Q}$
and $\hat{N} = \hat{R}$

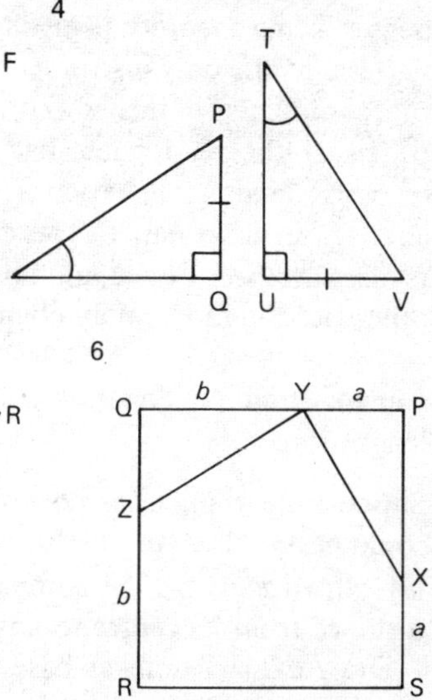

PQRS is a square. Are $\triangle$s PXY, QYZ congruent? If so, why ?

7 In the given triangle ABC, AB = AC. Line AD bisects angle BAC and meets BC at D. (This is the same diagram as Fig. 34, used to prove Theorem 7.) Show that the angles ADB and ADC are right angles.

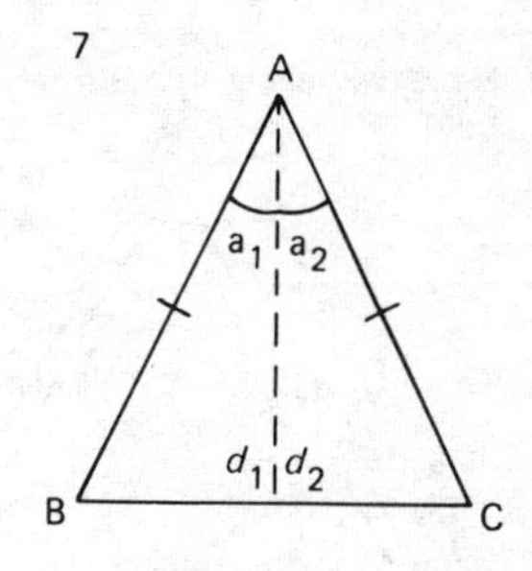

2 Some simple constructions using a compass

Many people find that at some time or other they need to draw carefully prepared diagrams. Quite apart from occupations in the fields of engineering, surveying, architecture and cartography, there are times in private life, such as planning with regard to improvements to one's home or garden, when ability to use simple mathematical instruments correctly can be a definite asset. This section, and the next, deal with some constructional aspects of geometry and, to a minor degree, some theoretical justification thereof.

Initially, the only instruments one needs are a pair of compasses, a ruler graduated in inches and/or centimetres, and a *nicely sharpened* pencil. If the reader has had little practice with compasses, it is recommended that he experiments in drawing circles and arcs (parts of circles) on scrap paper. The compasses should be held by the top projection (see Fig. 37a), not by either arm of the instrument, otherwise the radius may change.

Construction 1 To draw a triangle with all three sides of given lengths.

Suppose the triangle has sides of 5, 6 and 7 units respectively. (The unit could be, say, 1 cm *or* $\frac{1}{2}$ in. for convenient size.) Fig. 37a shows how to lay off a *radius* on the compasses, the radius of a circle being the distance from its centre to any point on its circumference (edge).

Draw BC = 7 units as base line. With centre B and radius 6 units draw an arc as shown. With centre C and radius 5 units draw an arc to cut the previous one at A. Join AB and AC, then traingle ABC is the one required (Fig 37b).

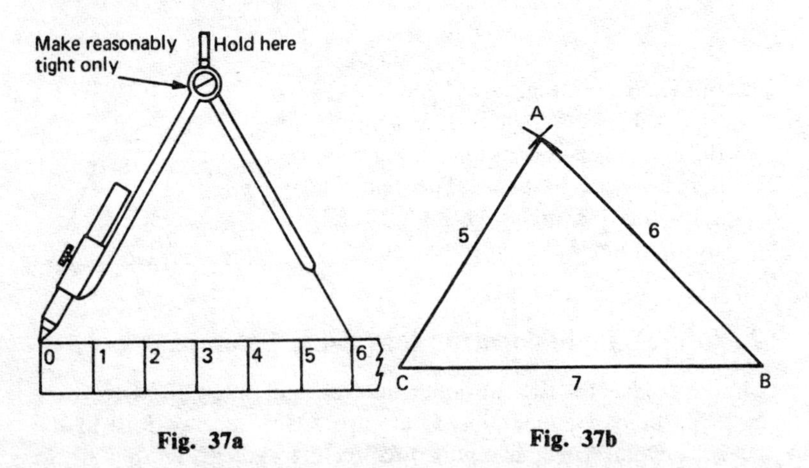

Fig. 37a **Fig. 37b**

It takes a little practice to gauge approximately where the point of intersection of two arcs (the point A, in this case) is likely to be, thereby keeping the arcs moderately short.

Construction 2 To bisect a given angle.

Let the given angle be A, with arms AB and AC (Fig. 38). With centre A and any convenient radius, draw equal arcs cutting AB and AC, at D and E respectively. With centres D and E in turn, draw equal arcs intersecting at F. Join AF, which is the required bisector of angle BAC. (N.B. AE = AD and EF = DF, but it is not necessary for the second pair to be of the same length as the first.)

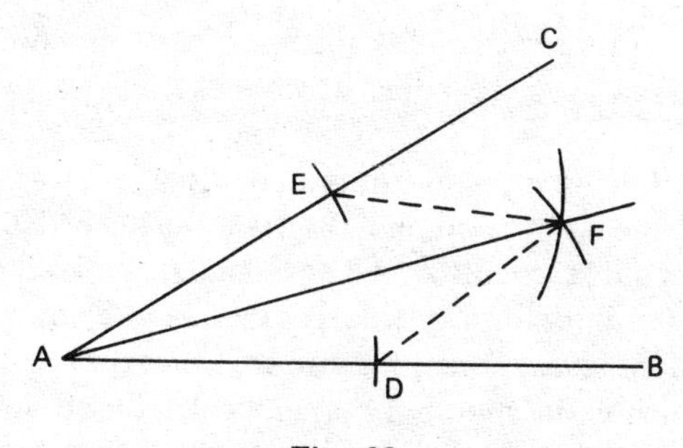

Fig. 38

(The proof is easy. If we join DF and EF, as shown by dotted lines, then in $\triangle$s ADF, AEF, we have AD = AE (equal arcs), DF = EF (equal arcs), and AF is common to both triangles.

$\therefore$ $\triangle ADF \equiv \triangle AEF$ (SSS) (Theorem 4)

$\therefore$ $D\hat{A}F = E\hat{A}F$.

Note that the dotted lines are *not* required in the actual construction.)

Construction 3 To bisect a given line (i.e. to find its midpoint by construction).

Let AB be the given line (Fig. 39). With centre A and radius rather more than $\frac{1}{2}$AB, draw arcs on both sides of AB. With centre B and the same radius, draw arcs cutting the first ones at C and D respectively. Lay the ruler along CD and after checking that the pencil accurately

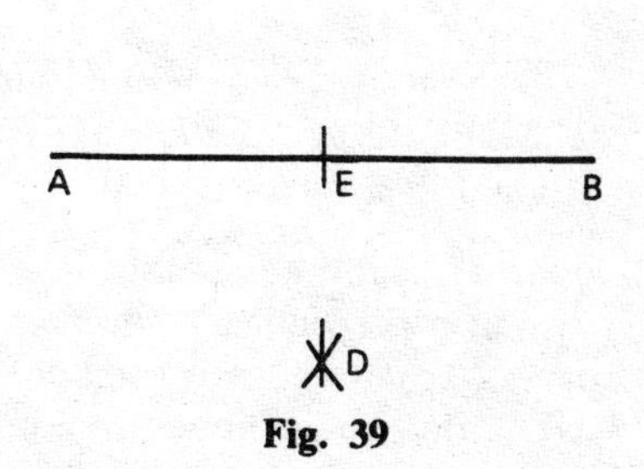

Fig. 39

passes through the intersections C and D, as shown, mark a small section of CD where it crosses AB at E. The point E bisects AB, i.e. AE = EB.

(The proof is straightforward but is appreciably longer than for construction 2. It is omitted.)

Construction 4 To construct a perpendicular (i.e. a line at right angles) to a given line at a given point on it.

There are two standard methods: the first is easier but does not always fit as conveniently on the paper as does the second, which is more compact.

Case (i): when the point does not present a snag. Let AB be the given line and P be the given point on it (Fig. 40). With centre P and suitable radius, draw equal arcs cutting AB at C and D, one on either side of P. With centre C and radius rather larger than before, draw an arc. With centre D and the same radius, draw an arc to cut the previous one at E. Join PE, which is the required perpendicular to AB at P.

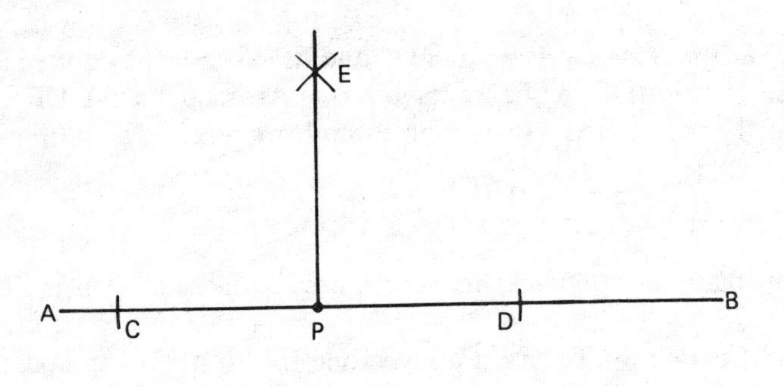

Fig. 40

(The proof is easy. Join CE and DE; one then shows that $\Delta CEP \equiv \Delta DEP$ (SSS), $\therefore C\hat{P}E = D\hat{P}E$, but $C\hat{P}E + D\hat{P}E = 180^\circ$ as CPD is a straight line; $\therefore C\hat{P}E = 90^\circ = D\hat{P}E$.)

Case (ii): near the end of the line or the edge of the paper. We take the case of a perpendicular needed at the end A of line AB, when BA cannot be conveniently extended more than a short distance (see Fig. 41). With centre A, draw an arc starting from P on AB and extending to

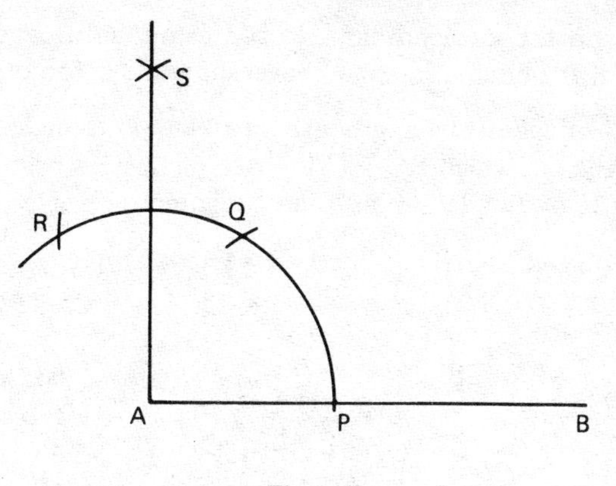

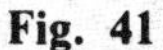
Fig. 41

between a quadrant (quarter circle) and a semi-circle in length. With centre P and same radius, mark off Q on the original arc. With centre Q repeat the process, thereby marking off R. With centres Q and R in turn, draw equal arcs intersecting at S. Join AS, which is the required perpendicular (i.e. angle BAS = 90°). The proof is omitted.

If we wish to construct an angle of 45°, we merely construct 90°, as in case (i) or (ii) above, and then bisect it as in Construction 2.

Definition (*22*) An equilateral triangle is one in which all the sides are equal.

Note: (i) The sides *opposite* the angles of a triangle, which are lettered in capitals, are often given the same letters, but in lower case (i.e. small type); (ii) It is very unwise to use lower case letters for sides

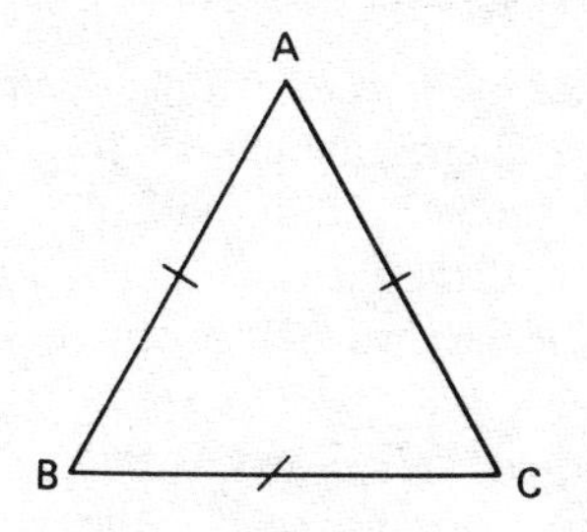

Fig. 42

and angles in the same problem, as confusion will frequently arise, especially if the circumflex sign $\hat{}$ over an angle is forgotten.

Theorem 8 Each of the angles of an equilateral triangle is 60°.

Draw ABC, in which BC = CA = AB, these sides being all of the same length, by the definition of an equilateral triangle (Fig. 42).

In $\triangle$ABC,	AC = AB $\therefore$	$\hat{B} = \hat{C}$	(Isosceles triangle: theorem 7)
Likewise	AB = CB $\therefore$	$\hat{C} = \hat{A}$	(same theorem)
Hence		$\hat{B} = \hat{C} = \hat{A}$	
But		$\hat{A} + \hat{B} + \hat{C} = 180°$	
$\therefore$		$3\hat{A} = 180° \Rightarrow \hat{A} = 60°$	
i.e.		$\mathbf{\hat{A} = \hat{B} = \hat{C} = 60°}$	

Construction 5 To construct an angle of 60°. (This follows at once from theorem 8.)

Draw a line BD and on it mark off BC = 3 cm, say. With centres B and C in turn, draw arcs each of radius 3 cm, intersecting at A. Join BA, then angle CBA = 60°.

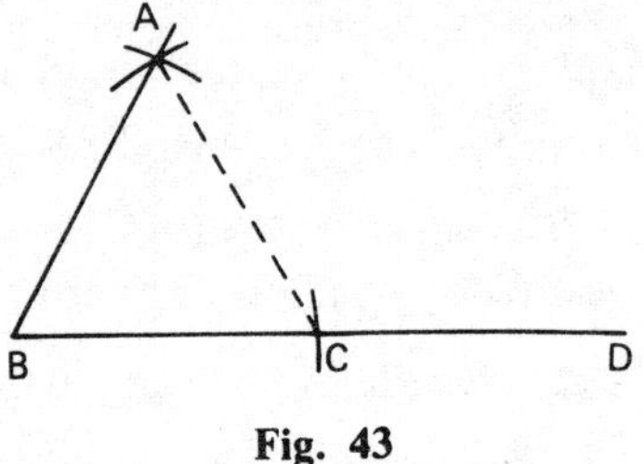

Fig. 43

(The proof is simple. Join AC, and we have an equilateral triangle!)

3 Some practical constructions using other instruments

We shall now make use of *set squares* and a *protractor*. Set squares are normally found in two shapes, namely 45° (Fig. 44b) and 30°/60° (Fig. 44d). The former is fundamentally half of a square bisected by a diagonal (Fig. 44a), and the latter is half of an equilateral triangle bisected by a perpendicular drawn from one vertex to the opposite side (Fig. 44c).

Although the set squares are shown as being of the same height here, for convenience, it is customary to find the 30°/60° set square to be taller and, of course, narrower than its matching 45° one. They used to

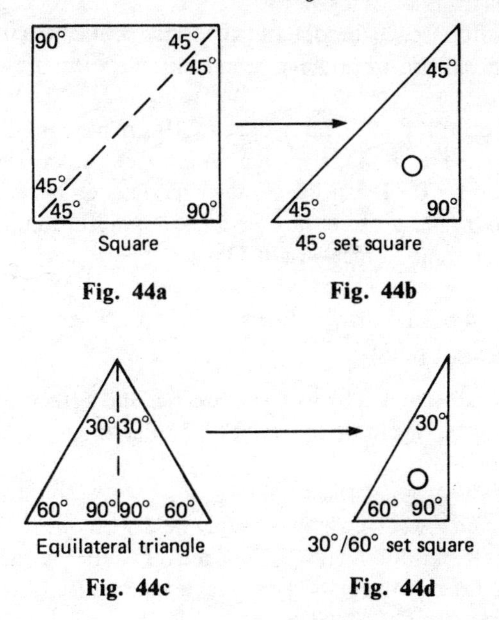

Fig. 44a **Fig. 44b**

Fig. 44c **Fig. 44d**

be made of wood or brass, but are nowadays generally of transparent plastic.

A protractor is an instrument for laying out angles on paper. It is usually constructed as a transparent plastic semi-circle, but it may be found in the form of a complete circle or an arc. (Sometimes one meets a hybrid instrument, consisting of two parts of a set square which have been specially severed and then joined by an adjustable arc calibrated as a protractor, but we shall not use one here.) Fig. 45 illustrates the

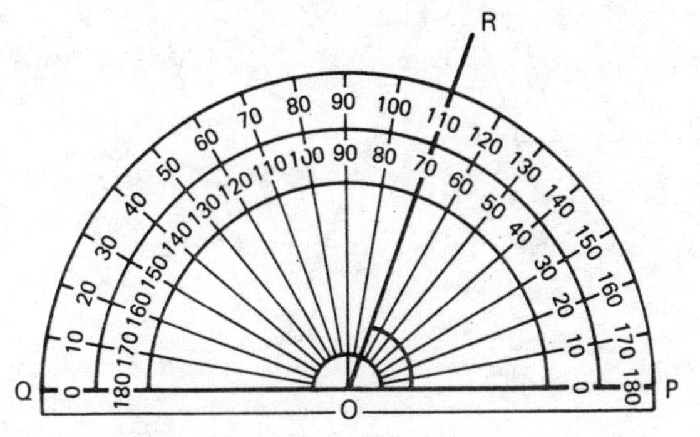

Fig. 45

standard semi-circular model, but the individual degree markings on the outermost semi-circle have been omitted in the drawing, for the sake of clarity.

O is the central point from which angles are measured, POQ is the *diameter* of the semi-circle; its length is twice the radius. The angle POR shown is 70°. Its supplementary angle is QOR, which is 110°, also shown on the protractor. One must be careful to take the actual angle needed, and to measure at O.

Construction 6 To draw a line parallel to a given line and to pass through a given point.

This can be constructed by using compasses and a ruler, but it is much easier and just as accurate to use *either* a ruler and a set square *or* two set squares.

Suppose the original line is AB (Fig. 46). Two set squares are aligned as shown. The lower one is then firmly held with one hand, while the upper one is carefully slid along the fixed one until the edge parallel to AB passes through the given point, which we shall call P. CD is then

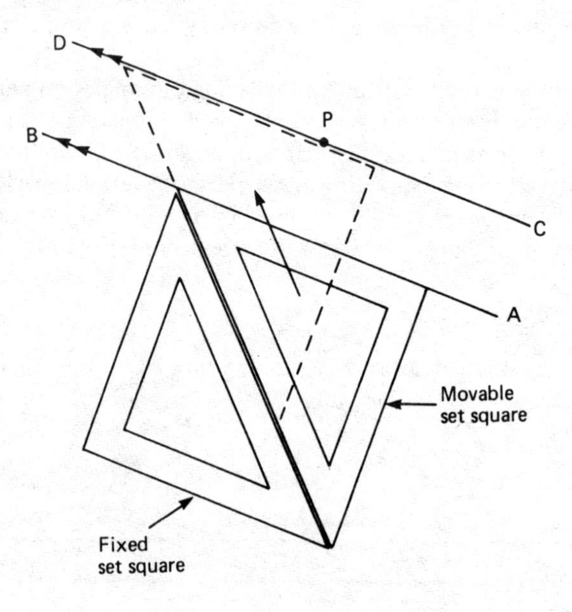

Fig. 46

drawn parallel AB. (If necessary, CD can be extended by the use of a ruler.)

CD, which passes through P, is clearly the line required.

Construction 7 The construction of a plane figure with one or more angles requiring the use of a protractor.

This is illustrated by an example which has been specially devised to warn the reader to watch carefully to see if more than one diagram may be drawn from given data.

Construct a quadrilateral ABCD in which angle A = 64°, AB = 3 units, AD = 2.5 units, BC = 2 units, CD = 1.5 units *and* angle C within the quadrilateral is *not* reflex.

In Fig. 47, lay off AB = 3 units and at A, using a protractor, mark off angle A = 64°. Draw AD = 2.5 units. With centre B, draw an arc of radius BC = 2 units; with centre D, draw an arc of radius DC = 1.5 units, cutting the previous arc at *two* points C_1 and C_2. Join BC_1 and DC_1; also join BC_2 and DC_2. We have *two* quadrilaterals ABC_1D and ABC_2D.

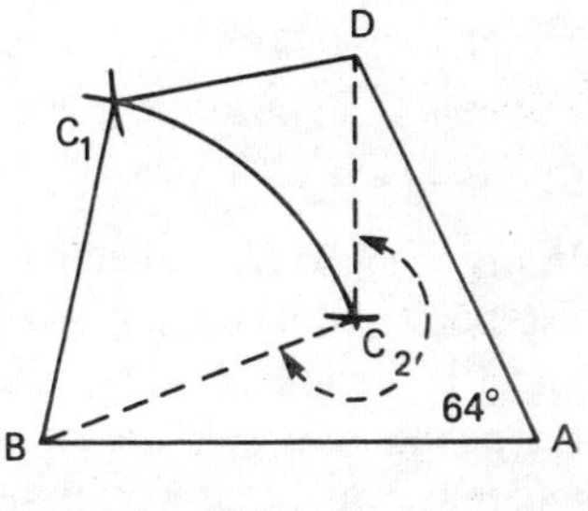

Fig. 47

Both could have been correct solutions if it were not for the fact that we are also instructed that angle C must not be reflex *within the quadrilateral.* Looking at Fig. 47, we see that angle BC_2D is reflex (marked dotted) and angle BC_1D is not. Hence ABC_1D is the correct quadrilateral.

We finish the chapter with one more construction. It is based on a property of the circle and it has not received earlier mention. We firstly give three simple definitions and state a theorem (without proof).

Definitions

23 The *circumference* of a circle may be defined as (a) the distance round its boundary line, (b) the boundary line itself.

Although these definitions are different, I do not recall an occasion when confusion occurred.

24 A *chord* is a straight line joining the ends of a circular arc.

25 A *diameter* is a chord which passes through the centre of a circle; it divides the circumference into two semi-circles.

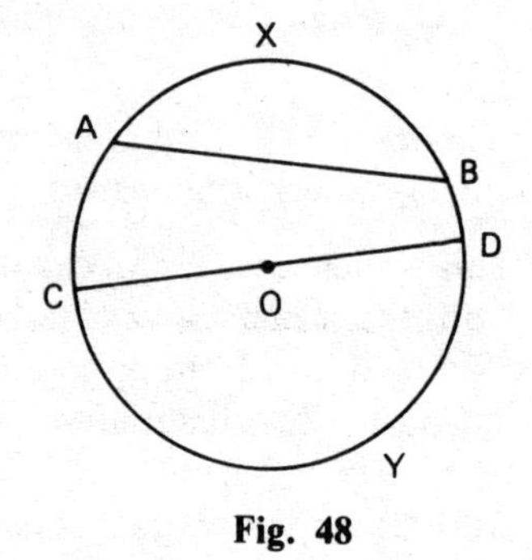

Fig. 48

In Fig. 48, the circumference is the whole way round the circle, i.e. AXBDYCA.

AB is a chord; it divides the circumference into a *minor* arc AXB (*less* than a semi-circle) and a *major* arc AYB (*greater* than a semi-circle).

CD is a diameter passing through O, the centre of the circle; the arcs CXD and CYD are both semi-circles.

Theorem 9 The angle in a semi-circle is a right angle. (The proof is omitted.)

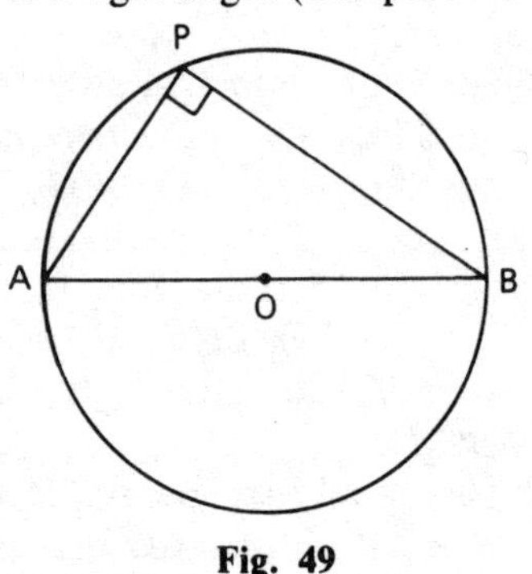

Fig. 49

In Fig. 49, AB is a diameter, and P is *any* point on the circumference. PA and PB are joined.

The theorem states that no matter where P is on the circumference (but A and B are special cases), then APB = 90°.

The converse of the theorem is also true, and it has practical applications. It reads as follows:

If A, P and B are points on the circumference of a circle, such that angle APB is a right angle, then AB is a diameter (which, of course, passes through the centre, O, in Fig. 49).

Construction 8 To find the centre of a circle, using a set square. (*Note*: All set squares normally have one angle of 90° on them.)

Using the set square, set up a right angle at P, say, and draw PR and PS cutting the circle at A and B respectively. Join AB. As angle APB is 90°, then AB is a diameter (Theorem 9, converse). See Fig. 50.

Repeat this using a point Q. We then obtain another diameter, CD in Fig. 50. As both diameters pass through the centre of the circle (by definition of a diameter), their point of intersection, O, is the centre

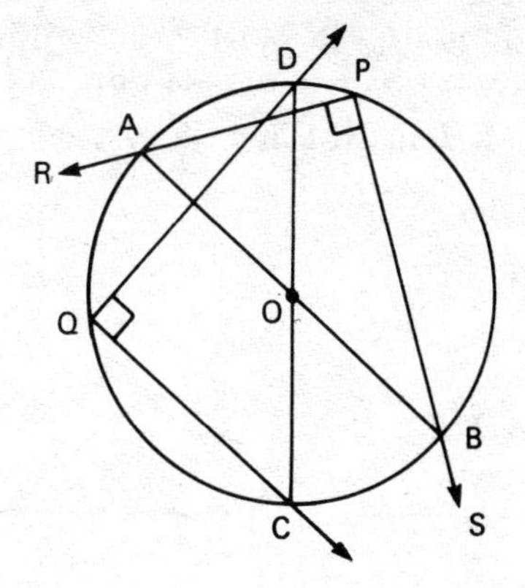

Fig. 50

required. (*Note*: the angle of interesection of AB and CD should be reasonably large; say, not less than 40°.)

There are other methods, using compasses, but they are not always convenient. If, for example, one needed to find the centre of a circular metal plate, the above method, carefully applied, should be accurate enough.

EXERCISE 2

In questions 1 to 5, use a ruler, compasses and good pencil only (although a rubber is allowed!).

1 Construct a triangle with sides 2.5, 3.0 and 3.4 cm, respectively.

2 AB is a line of length 2 in (*or* 5 cm). At A construct a perpendicular to AB. Hence construct the square ABCD. Measure the diagonals AC and BD.

3 PQ is a line of length 6 cm. At P, construct a line PS, of length 4 cm, such that at angle QPS = 60°. Hence construct parallelogram PQRS.

Join SQ. Measure SQ and angle SQR.

4 (*Construction* 9) Although this is a standard construction using ruler and compasses, and as such is worth remembering, it makes a useful exercise.

Draw a line FG and take a point H about 2.5 cm away from it. Construct a perpendicular from H to FG, as follows. With centre H and a suitable radius, draw short arcs cutting FG at K and L respectively. With centres K and L in turn, using a new radius rather shorter than before, but the same for K and L, draw two arcs intersecting at M. Join HM and, in case (a), produce to meet FG at N.

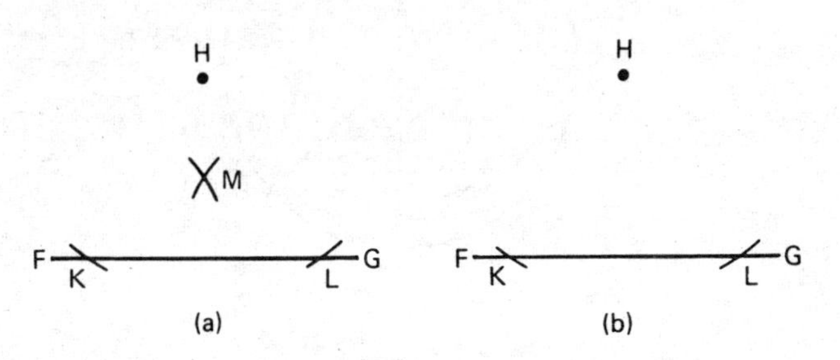

Case (a), M on the *same* side of FG as is H, has been nearly completed. Case (b), M on the *opposite* side of FG from H, is just outlined. The reader is advised to draw both cases for practice.

The choice of (a) or (b) often depends on how much space is available on the paper: (a) needs less space; (b) is slightly more accurate.

5 Construct a right-angled triangle in which the hypotenuse is 6.5 cm and one other side is 2.5 cm. What is the length of the third side?

6 Construct a trapezium ABCD in which AB $= 5$ cm, angle B $= 63°$, BC $= 3$ cm, CD $= 5.5$ cm, and AD $\|$ BC. Measure AD and angle C. (*Notes.* (i) Set square and protractor are needed; (ii) There is only one correct case for the given data.)

7 Find a circular metal or wooden object of reasonable size, at home. Locate the centre using the method of construction 8. (*Alternatively*, place a circular object on a piece of paper. Draw round it and then find the centre of the drawing.)

7

Mensuration of Rectilinear Figures

1 A definition of mensuration

For most people who are familiar with the word, mensuration calls to mind a strictly practical aspect of mathematics. This is a very reasonable view, for mensuration is the process of finding, by measurement and calculation, the lengths, areas, volumes or what you will, of plane figures, curves and shapes, or of three-dimensional curves, surfaces and solids. Examples of these, in corresponding order, are a line, a circular arc, a rectangle, a spiral spring, a pyramid, a cylinder and a sphere – note that the pyramid, cylinder and sphere can be hollow or solid.

Before reaching this stage, however, some theoretical knowledge is usually of value, in the sense of knowing, or having the ability to construct, a formula in which to substitute numerical values. The formula may be simple, such as arises in finding the area of a triangle (for which the theory is given – see below), or it may be slightly more advanced, as in the case of obtaining the volume of a sphere by means of calculus. The theory of the latter will not be included.

In this chapter we shall consider *rectilinear* plane and solid figures (i.e. ones bounded by straight lines). Areas and volumes will often be expressed as, say,

$$\triangle \mathrm{ABC} = \tfrac{1}{2}xy, \text{ Rect. PQRS} = hk, \text{ volume of pyramid} = Ah,$$

wherein the sign = means 'is equal in area to'. This is a slightly different meaning from that adopted in most parts of the book, but in practice it causes no problem.

2 Plane figures bounded by straight lines

(i) **The area of a rectangle** is the product of its length and its breadth, i.e. length × breadth. It is often convenient to think of this as the product of its base and its height, where we refer to the *length* as *base* and the *breadth* as *height*. Two specific examples of these alternative names are (a) when considering the area of a triangle, (ii) when measuring the area of a wall in a room (Fig. 51).

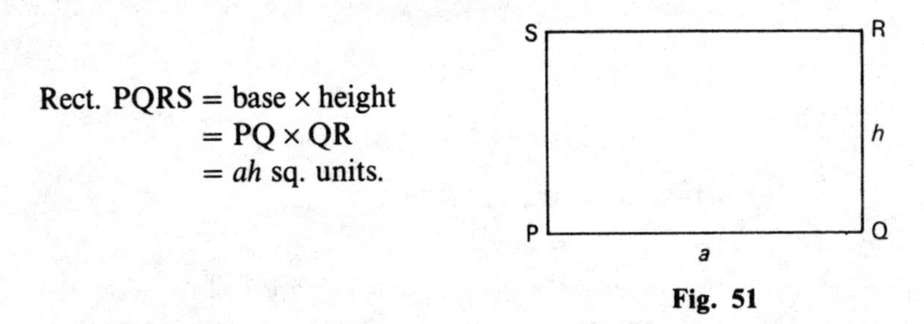

Rect. PQRS = base × height
$= PQ \times QR$
$= ah$ sq. units.

Fig. 51

(ii) **The area of a square** is the square of the length of any side, as all the sides are equal.

In Fig. 52, if each side is of length c units,

$$\text{Sq. KLMN} = KL^2$$
$$= c^2 \text{ sq. units.}$$

As mentioned earlier, a square is a particular case of a rectangle.

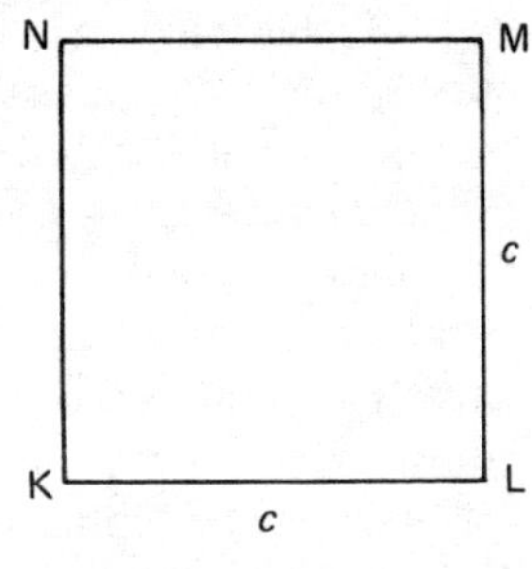

Fig. 52

(iii) **The area of a triangle** is one-half of the product of its base and its corresponding perpendicular height, i.e. *it is half the area of a rectangle on the same base and of the same height as the triangle.*

Any one of the three sides of a triangle may be taken as base; the perpendicular height is then measured from the vertex opposite to the chosen base.

Fig. 53a illustrates an *acute-angled* triangle (i.e. one in which all the angles are less than 90°) with its associated rectangle on the same base and of the same height.

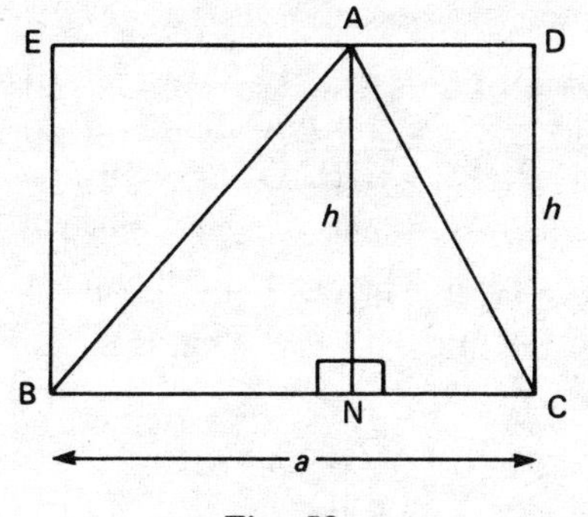

Fig. 53a

The proof is given in outline. It is based on the fact that a diagonal of a rectangle bisects it (in fact, divides it into two congruent triangles).

From Fig. 53a

$$\text{rect. DEBC} = \text{rect. DANC} + \text{rect. AEBN}$$
$$= 2 \times \triangle\text{ANC} + 2 \times \triangle\text{ANB} = 2 \times \triangle\text{ABC}$$
$$\therefore \quad \text{ABC} = \tfrac{1}{2}\text{rect.DEBC} = \tfrac{1}{2}\text{BC} \times \text{CD} = \tfrac{1}{2}\text{BC} \times \text{AN}$$
$$= \tfrac{1}{2}ah, \text{ i.e. } \tfrac{1}{2}\text{base} \times \text{height}.$$

Fig. 53b shows a similar picture for an *obtuse-angled* triangle (i.e. one in which one angle lies between 90° and 180°). This is a little more tricky, because in a triangle of this shape, the perpendicular to a chosen side *may* fall outside the triangle. In two cases, it does (one such is shown); in one case it does not, and Fig. 53a then applies unchanged!

The lettering in Fig. 53b corresponds exactly with that of Fig. 53a, although the diagrams themselves differ noticeably.

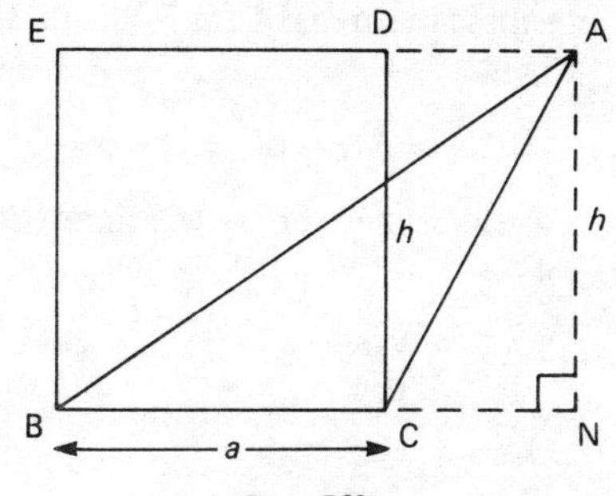

Fig. 53b

From Fig. 53b

$$\text{Rect. DEBC} = \text{rect. AEBN} - \text{rect. ADCN}$$
$$= 2 \times \triangle\text{ANB} - 2 \times \triangle\text{ANC} = 2 \times \triangle\text{ABC}$$
$$\therefore \quad \text{ABC} = \tfrac{1}{2}\text{rect. DEBC} = \tfrac{1}{2}\text{BC} \times \text{CD} = \tfrac{1}{2}\text{BC} \times \text{NA}$$
$$= \tfrac{1}{2}ah, \text{ i.e. } \tfrac{1}{2}\text{base} \times \text{height.}$$

Example 1 In $\triangle$PQR, QR = 5.6 cm and the length of the perpendicular from P to QR is 4.5 cm. What is the area of the triangle?

In the diagram,

$$\triangle\text{PQR} = \tfrac{1}{2}\text{QR} \times \text{PN} = \tfrac{1}{2} \times 5.6 \times 4.5 \text{ cm}^2$$
$$= \mathbf{12.6\ cm^2}$$

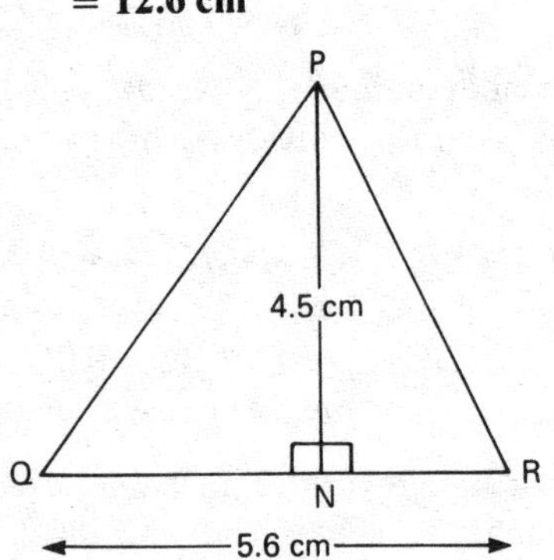

Observe that we do not need to know the shape of the triangle when using base and corresponding height.

Example 2 In the given figure, ABCD is a rectangle in which AB = 7 cm and BC = 5 cm. Find the area of the rectangle.

P and Q are any two points which lie, respectively, on DC and on DC produced. Determine the areas of $\triangle$s PAB and QAB. Hence show that $\triangle$PAB = $\triangle$QAB.

$$\text{Rect. ABCD} = \text{AB} \times \text{BC} = 7 \times 5 = \mathbf{35\ cm^2}$$

But from above, as the triangles have the same height and the same base as the rectangle.

$$\therefore \quad \triangle\text{PAB} = \tfrac{1}{2}\text{ rect. ABCD} = \tfrac{1}{2} \times 7 \times 5$$
$$= \mathbf{17.5\ cm^2}$$

and $\quad \triangle\text{QAB} = \mathbf{17.5\ cm^2}$

$$\therefore \quad \triangle\text{PAB} = \triangle\text{QAB.}$$

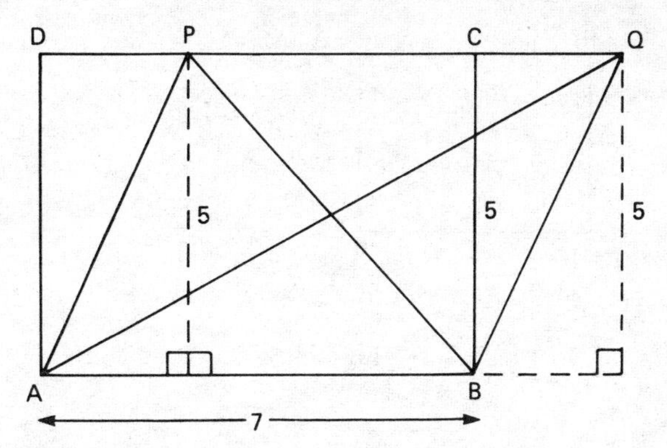

Incidentally, the fact that the triangles have the same area should come as no surprise, as each is $\frac{1}{2} \times$ base $\times$ height of the same rectangle. Furthermore, it does not matter where P and Q are, as long as they are on DC *or* DC produced to the right *or* CD produced to the left. The selection of P within DC and Q beyond C was merely to illustrate the two cases dealt with in Figs. 53a and 53b.

(iv) **The area of a parallelogram** is equal to the area of a rectangle on the same base and between the same parallels, i.e. the area is base times *perpendicular* height. (*Note*: ∥ gm is an abbreviation for the word 'parallelogram'; quad. stands for 'quadrilateral'.)

A brief outline of the proof is given below, but it may be omitted without affecting subsequent work. Case (ii) requires care (Fig. 54).

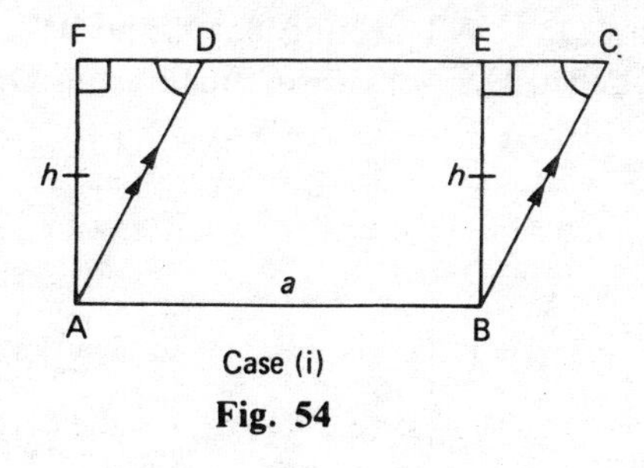

Case (i)

Fig. 54

$$\triangle BCE \equiv \triangle ADF \text{ (AA Corr. S)}$$

$$\begin{aligned} \therefore \parallel \text{gm ABCD} &= \text{quad. ABED} + \triangle BCE \\ &= \text{quad. ABED} + \triangle ADF \\ &= \text{rect. ABEF} \\ &= ah \text{ sq. units} \end{aligned}$$

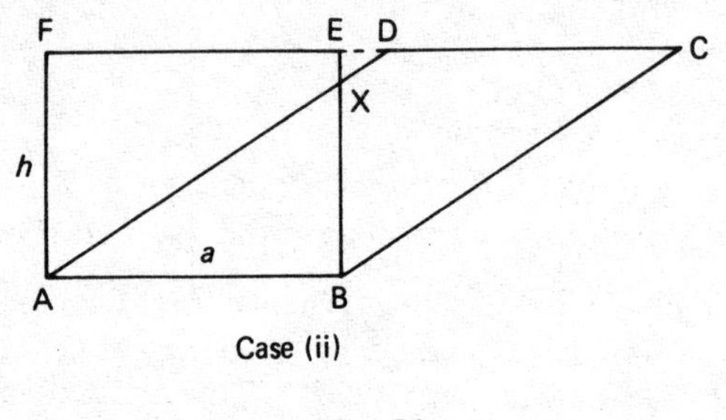

Case (ii)

Fig. 54

$$\begin{aligned}\triangle BCE &\equiv \triangle ADF \text{ (as before)}\\ \therefore \| \text{gm } ABCD &= \triangle ABX + \triangle BCE - \triangle DEX\\ &= \triangle ABX + \triangle ADF - \triangle DEX\\ &= \triangle ABX + \text{quad. } AXEF\\ &= \text{rect. } ABEF\\ &= ah \text{ sq. units}\end{aligned}$$

(v) **The area of a trapezium** is half the sum of the parallel sides multiplied by the distance between them. (*Note*: The symbol $\perp$ means 'perpendicular to'.)

The proof is easier than for the parallelogram. The parallel sides are of lengths a and b respectively.

Let the perpendicular height (i.e. the distance between the parallel sides) be p. Draw DX $\perp$ AB meeting it at X, and BY $\perp$ DC produced meeting it at Y, then DX = YB = p. Finally, join DB (Fig. 55).

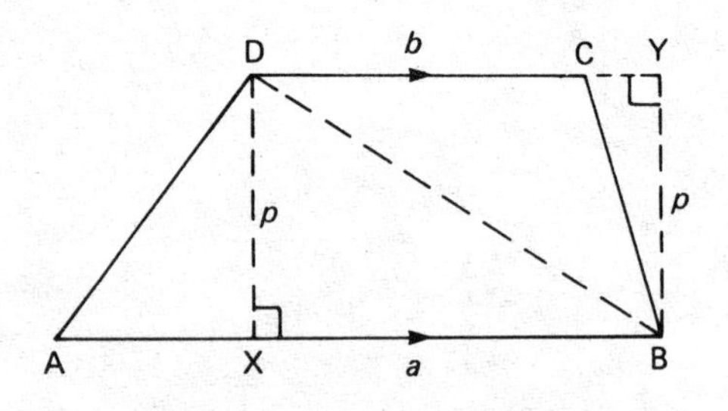

Fig. 55

We have

$$\text{trap. ABCD} = \triangle\text{ABD} + \triangle\text{DCB}$$
$$= \tfrac{1}{2}\text{AB} \times \text{DX} + \tfrac{1}{2}\text{DC} \times \text{YB}$$
$$= \tfrac{1}{2}ap + \tfrac{1}{2}bp$$
$$= \tfrac{1}{2}(a+b)p \text{ sq.units, which is the required result.}$$

Example 3 A piece of land is in the form of a trapezium, the parallel sides being of length 63 yd and 52 yd, respectively. The distance between them, measured perpendicularly, is 44 yd. Find the area, expressed as a decimal correct to two places, (i) in acres (ac), (ii) in hectares (Ha), given that 1 Ha $\simeq$ 2.47 ac.

The area required is

$$\tfrac{1}{2}(63+52) \times 44 \text{ sq. yd}$$
$$= 115 \times 22 \text{ sq. yd}$$
$$= \frac{115 \times 22}{4840} \text{ ac}$$
$$\simeq 0.523 \simeq \mathbf{0.52\ ac} \text{ (2 dec. pl.)}$$
$$\simeq \frac{0.523}{2.47} \simeq 0.211 \simeq \mathbf{0.21\ Ha} \text{ (2 dec. pl.)}$$

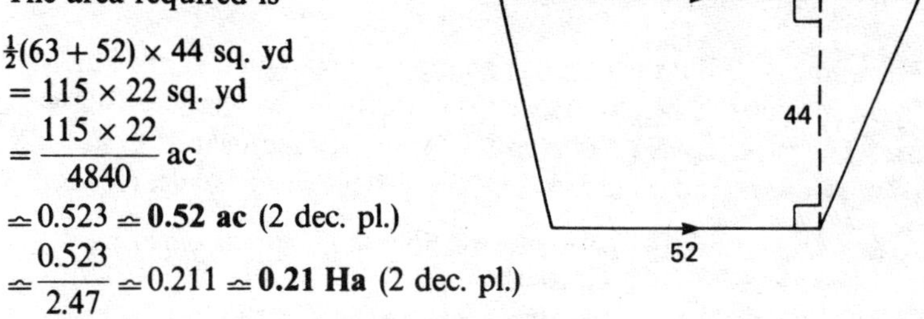

Note that to get the area in hectares, correct to two decimal places, we need at least three places in the working.

3 Wallpapering a room

Definition The perimeter of a plane figure is the total distance round its edge(s).

Suppose we wish to paper a rectangular room, of length a metres and breadth b metres, to a height of h metres. The easiest way to look at the problem is to imagine the room to be cut open and laid out flat (Fig. 56).

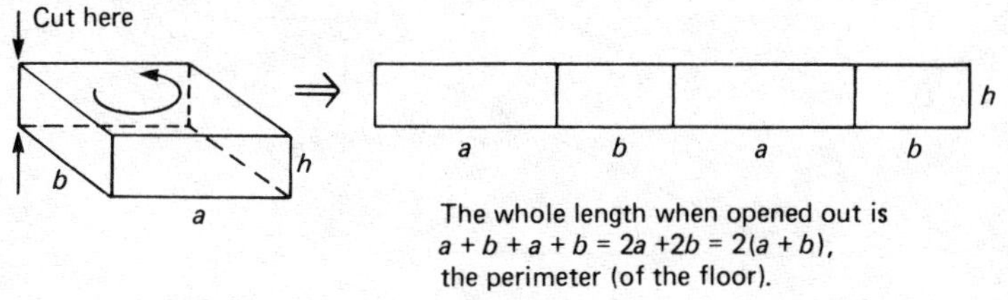

The whole length when opened out is $a + b + a + b = 2a + 2b = 2(a + b)$, the perimeter (of the floor).

Fig. 56

$\therefore$ the area to be prepared = perimeter of floor × height of wall to be papered

$$= 2(a+b)h = 2h(a+b), \text{ which is neater.}$$

We usually take the doors and windows to be *included* in the papering. This means that we shall have a little extra to allow for wastage but, unless the windows are exceptionally large and/or the doors numerous, there may still not be enough paper. Pattern matching also can be very wasteful and a minimum of 10% extra, or one roll more than is calculated to cover the measured area, is strongly advised for rooms of average size. Remember, also, that ordering extra rolls at a later date may produce another snag – there may be a difference in the shade of printing!

Example 4 Find the cost of papering a rectangular room 5 m long by 3.5 m wide to a height of 2 m. (The skirting board has been allowed for.) The paper costs £4.80 a roll, which is 10 m long and 50 cm wide. (Ignore doors and windows and allow 10% for wastage, etc.)

Area to be covered = perimeter of floor × height of papering

$$= 2h(a+b)$$
$$= 2 \times 2(5+3.5) = 4 \times 8.5 = 34 \text{ m}^2$$
$$(\text{m}^2 = \text{square metres})$$

Area of one roll of paper = 10 m × 50 cm = $10 \times \frac{1}{2}\text{m}^2 = 5\text{ m}^2$
$\therefore$ the number of rolls required is $34 \div 5 = 6.8$
We need to add 10%, i.e. $\frac{1}{10}$th, to cover wastage, etc., giving $6.8 + 0.68 = 7.48$ rolls, but one cannot buy part rolls. The total number for safety is therefore 8 rolls.

The cost is £4.80 × 8 = **£38.40**

Example 5 Using Pythagoras' theorem, calculate the area of

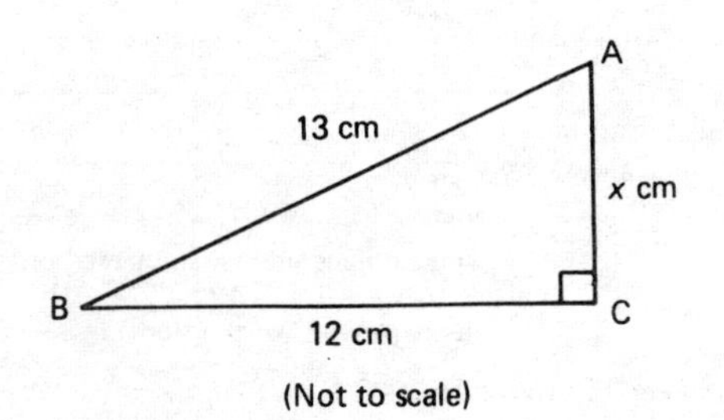

(Not to scale)

$\triangle$ABC, in which angle C = 90°, AB = 13 cm and BC = 12 cm. (See page 7.)

We firstly draw the diagram, and let AC = x cm.

From Pythagoras' theorem, $\quad AC^2 + BC^2 = AB^2$

i.e. $\quad x^2 + 144 = 169$

giving $\quad x^2 = 169 - 155 = 25$

$\therefore \quad x = \sqrt{25} = 5$, i.e. AC = 5 cm.

$\therefore \quad$ area of $\triangle ABC = \frac{1}{2}BC \times AC$

$= \frac{1}{2} \times 12 \times 5\,\text{cm}^2 = \mathbf{30\,cm^2}$

EXERCISE 1

Find the areas of the triangles ABC in questions 1–5; in the cases of 3 and 4, also find the perimeters.

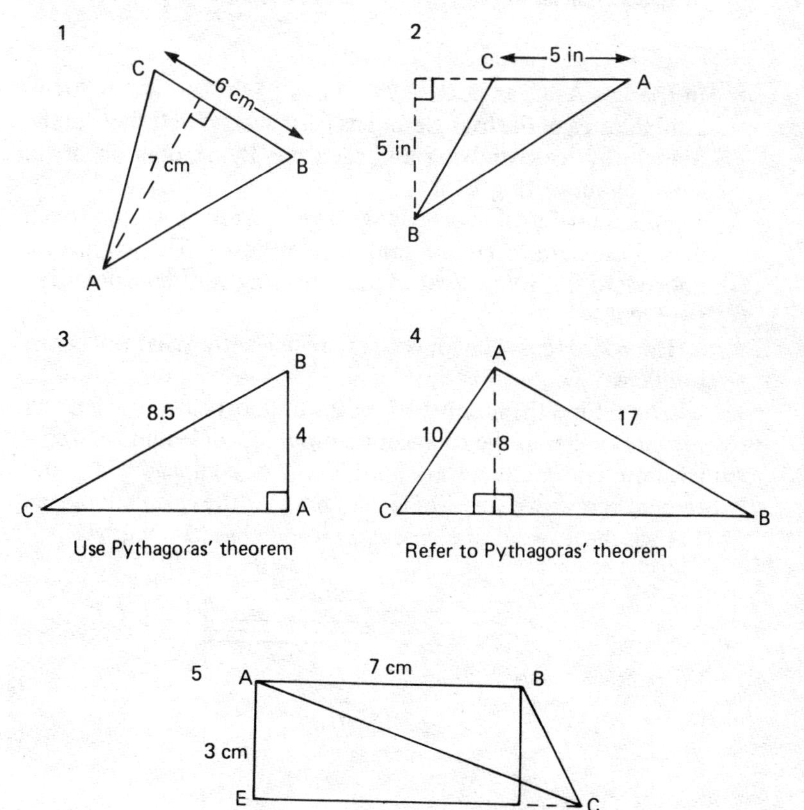

Use Pythagoras' theorem

Refer to Pythagoras' theorem

ABDE is a rectangle; C is on ED produced

Find the areas and perimeters of the figures in questions 6 and 7. (In 7, construction lines are needed, as shown in 6.)

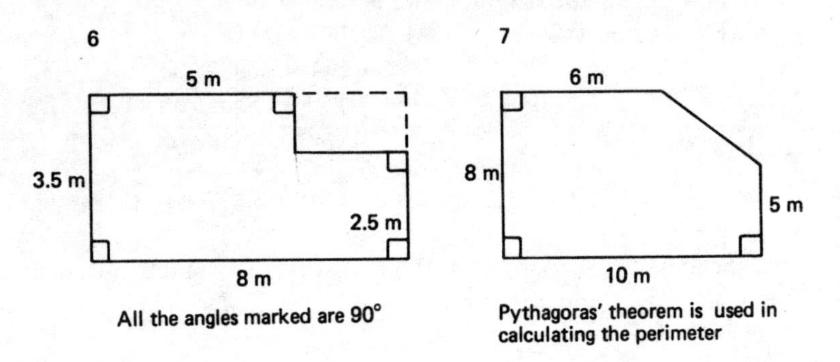

All the angles marked are 90°

Pythagoras' theorem is used in calculating the perimeter

8 In triangle ABC, angle A = 90°, AB = 25 cm and AC = 20 cm. Find the area of the triangle and its perimeter. (Firstly lay out the triangle by scale drawing and then use Pythagoras' theorem; check by measuring BC.)

9 The diagram illustrates a rectangular room, 6 m by 4 m, on which floorboards are to be laid longways on joists. The width of a floorboard is 15 cm. What is the *total* length of floorboarding required?

The boards cost 80p for each metre in length. What will be the total cost?

(*Hint*: Find the number of widths of floorboards needed and remember that, if the total comes out as a whole number and a fraction, one needs to buy floorboards of complete width and then to saw down one whole length; e.g. if the calculation were $17\frac{1}{3}$ widths, it would be necessary to purchase 18 widths.)

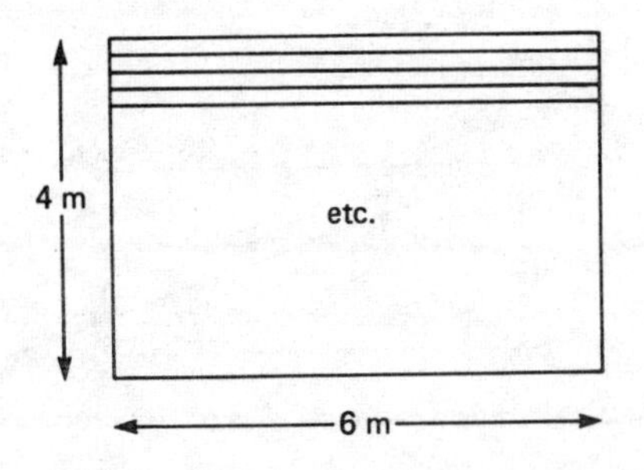

10 The figure represents the plan of a room with a symmetrical bay window and an inset chimney breast. Find the area of the ceiling, in square metres, correct to one decimal place. Take the ceiling to have the same area as the floor. All angles are right-angles except for the bay window.

(*Hints*: (i) The plan needs to be subdivided into simple shapes as in nos. 6 and 7 above; (ii) Remember that expressions such as $0.4 \times 0.9 = 0.36$, not 3.6.)

The ceiling is to be given two coats of emulsion. Assuming that 5 litres of emulsion will cover $60\,\text{m}^2$ with one coat, find the number of litres required, correct to the nearest 0.5 litre.

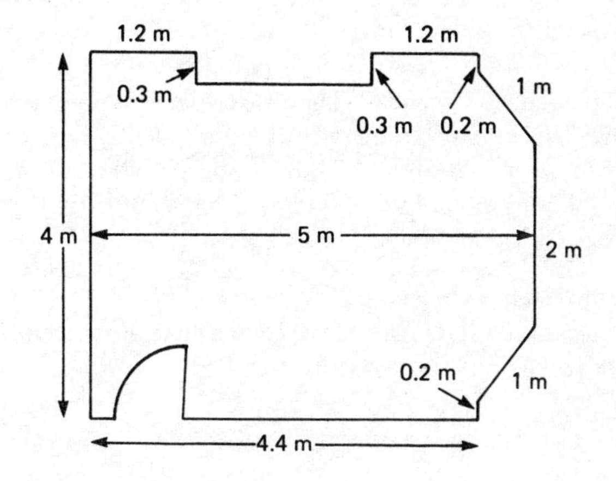

11 *Surveying of Land*. In the surveying of a polygonal area of land, such as is illustrated below, it is customary to take a principal line (the long diagonal AD in our picture) from which distances are measured perpendicularly to other vertices (B, C and E in our case). We thus break down the area into triangles and trapeziums.

In the given field ABCDE, the measurements are shown in metres, the long diagonal AD being of length 268 m. Find the area of the field in hectares, correct to 2 decimal places. (Note that BCRP is a trapezium, for BP and CR are parallel, both being perpendicular to AD; 1 hectare (Ha) $= 10\,000\,\text{m}^2$.)

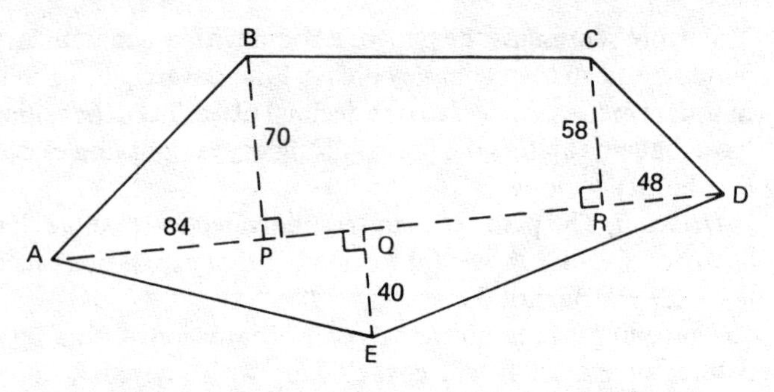

4 Three-dimensional rectilinear figures (polyhedra)

A polyhedron is a many-sided, three-dimensional figure bounded by plane faces, which of course can only meet along straight lines. The simplest such figure is a *cube*. But before examining some of its properties, we are confronted by a paradox with regard to three-dimensional figures in general. They are called *solids*, but a solid may be hollow! This kind of problem is all too frequently met when using words with several meanings. Thus, if any possible confusion may arise, we shall say either 'hollow cube' or 'solid cube', and use similar expressions for other 'solids'.

Some properties of the cube

In Fig. 57a, ABCDGHIJ is a cube of edge a units. For convenience we shall take these units to be centimetres (cm).

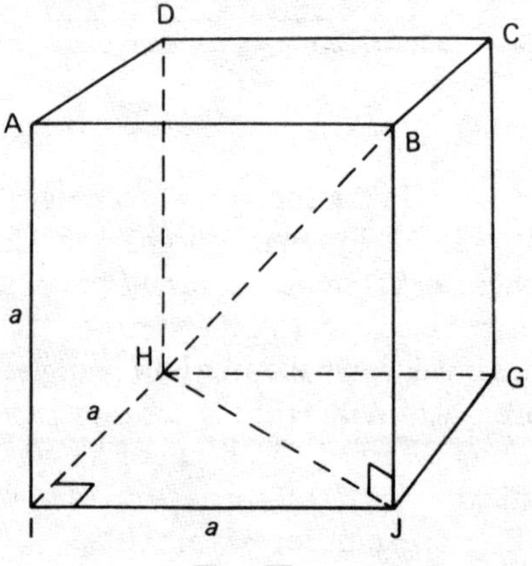

Fig. 57a

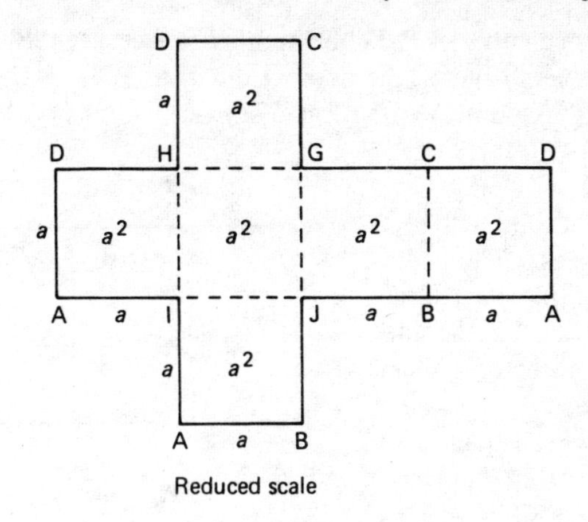

Fig. 57b

The volume of a *solid* cube is a^3 (cm^3)

The surface consists of six identical squares. See Fig. 57b, which is the *net* of Fig. 57a, but on a smaller scale; such a net is a flat (plane) area which can be folded to make the *hollow* cube itself. The figure shows very clearly how the cut edges, such as HD and HD, are joined together to make the cube. It is easy to follow the lettering in the cube and in the corresponding net.

The *total surface area* is S, where $S = 6 \times$ sq. ABCD $= \mathbf{6a^2}$ (cm^2)

In Fig. 57a, HJ and HB are joined, where HJ is a diagonal of a *face*, GHIJ, and HB is a main diagonal of the cube (from one vertex to its diagonally opposite vertex).

By Pythagoras' theorem, $\text{HJ}^2 = \text{HI}^2 + \text{IJ}^2$ (as angle HIJ $= 90°$)

$$= a^2 + a^2 = 2a^2$$

Again as JB is perpendicular to plane GHIJ it is perpendicular to every line in it,

$$\therefore \qquad \text{JB} \perp \text{JH}$$

So by Pythagoras' theorem, $\text{HB}^2 = \text{HJ}^2 + \text{JB}^2$ (as angle HFB $= 90°$)

$$= 2a^2 + a^2 = 3a^2$$

$\therefore$ the length of a main diagonal of a cube is $\sqrt{3}\boldsymbol{a}$ (cm) (sometimes written $a\sqrt{3}$).

Question In a cube, let F_n be the number of faces, V_n the number of vertices (corners), and E_n the number of edges. What is the value of $F_n + V_n - E_n$? (Count them, in Fig. 57a.)

The cuboid

A cuboid is a solid with rectangular faces. The commonest example must surely be a plain brick. Let the sides of the cuboid be a, b, c cm. (In Fig. 58a,
$AB = DC = HG = IJ = a$,
$BC = AD = IH = JG = b$, and
$AI = BJ = CG = DH = c$).

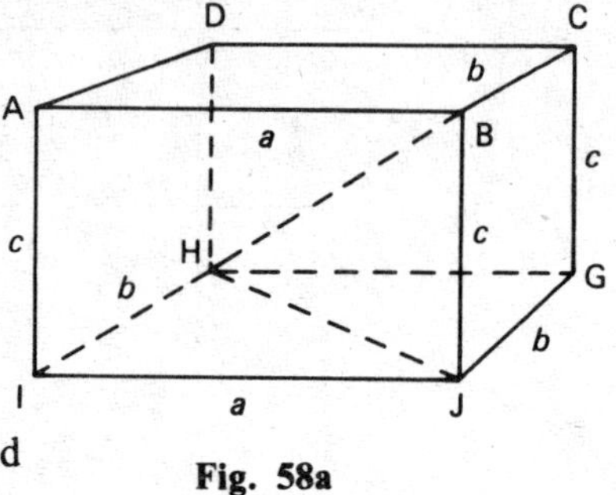

Fig. 58a

The volume of the cuboid is $\boldsymbol{abc}$ (cm³)

Also $HJ^2 = IJ^2 + HI^2 = a^2 + b^2$ (Pythagoras' theorem)
and so $HB^2 = HJ^2 + JB^2 = (a^2 + b^2) + c^2 = a^2 + b^2 + c^2$
$\therefore$ each main diagonal (AG, BH, CI, DJ) is of length $\sqrt{a^2 + b^2 + c^2}$ **cm**

The surface consists of three pairs of rectangles, as can clearly be seen on the net. The areas are marked in each rectangle (Fig. 58b).
$\therefore$ total surface area is **2(*bc*+ *ca*+ *ab*)** (cm²)

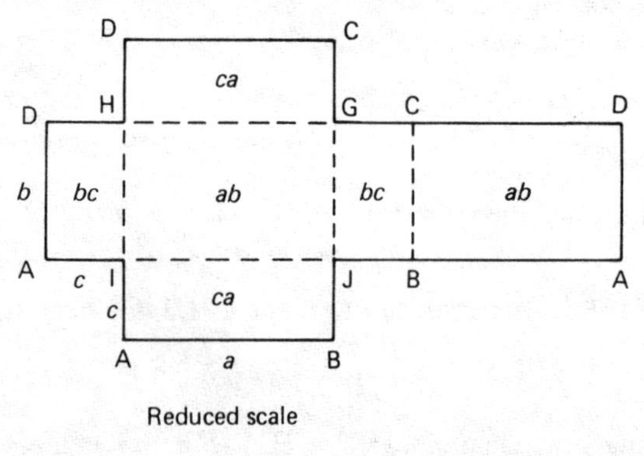

Fig. 58b

Example 6 A large rectangular box is of length 2.4 m, breadth 80 cm and height 60 cm. Find, in cubic metres, the volume of space occupied by the box.

The box and lid, *excluding* the base of the box, are to be painted externally. Calculate the number of square metres to be covered.

A half-litre (500 ml) can of paint states on the outside 'sufficient to cover 6 to 7 square metres'. Is this adequate for the job?

Determine the length of a main diagonal of the box.

The volume (V) occupied by the box is given by $V = abc \text{ m}^3$, where $a = 2.4 \text{ m}$, $b = 80 \text{ cm} = 0.8 \text{ m}$, and $c = 60 \text{ cm} = 0.6 \text{ m}$.

$\therefore$ The volume is $2.4 \times 0.8 \times 0.6 \text{ m}^3$
$= \mathbf{1.152 \text{ m}^3}$

The area to be painted (A) is $ab + 2bc + 2ca \text{ m}^2$, as the base (of area ab) is omitted.

$$\therefore \quad \text{A} = ab + 2bc + 2ca = 2.4 \times 0.8 + 2 \times 0.8 \times 0.6 + 2 \times 0.6 \times 2.4 \text{ m}^2$$
$$= \mathbf{5.76 \text{ m}^2}$$

The can of paint is therefore just adequate for the work, if one is not too generous in application.

The diagonal of the box (D) is given by

$$D^2 = a^2 + b^2 + c^2$$
$$= (2.4)^2 + (0.8)^2 + (0.6)^2 \text{ m}^2 = 5.76 + 0.64 + 0.36 \text{ m}^2 = 6.76 \text{ m}^2$$
$$D = \sqrt{a^2 + b^2 + c^2} = \sqrt{6.76} \text{ m} = \mathbf{2.6 \text{ m}} \text{ (exactly)}.$$

Example 7 The external measurements of a rectangular box (including the lid) are 1.5 m, 1.2 m and 0.4 m, these being the length, breadth and height, respectively. The wood of which the box is made has a uniform thickness of 2 cm. Find the internal capacity of the box, in cubic metres.

Express this capacity as a percentage of the external volume, correct to one decimal place.

We must remember that the length has *two* thicknesses of wood, one at each end. The breadth and height have the same. Hence the *internal* measurements are, in this order,

$$a = 1.5 - 0.04 \text{ m} = 1.46 \text{ m}; \quad b = 1.2 - 0.04 \text{ m} = 1.16 \text{ m};$$
$$c = 0.4 - 0.04 \text{ m} = 0.36 \text{ m}.$$

$\therefore$ the internal capacity of the box is

$$1.46 \times 1.16 \times 0.36 \text{ m}^3$$
$$\simeq 0.60969 \text{ m}^3 \simeq \mathbf{0.6097 \text{ m}^3}$$

Also the external volume of the box is $1.5 \times 1.2 \times 0.4\,\text{m}^3 = 0.72\,\text{m}^3$,
$\therefore$ the internal capacity expressed as a percentage of the external volume is approximately

$$\frac{0.6097}{0.72} \times 100 \simeq 84.68 \simeq \mathbf{84.7\%}$$

The right pyramid on a square base

This is just the kind of thing to have brought a warm glow to any ancient Egyptian. We are not, even so, studying this solid for the above reason, but because it is a useful shape for understanding a little about mensuration of polyhedra.

A polyhedron of the type named in the heading (Fig. 59a) has, in addition to its square base ABCD, four congruent triangles, VAB, VBC, VCD and VDA, as faces. These faces have a common vertex V, situated directly over the centre of the base ABCD. This means that the EV is perpendicular to ABCD, i.e. it is perpendicular to any line in the plane containing ABCD.

We firstly take a simple case in which the four triangles named above are equilateral. Let an edge be of length $2a$, then this is the length of *all* the eight edges of the pyramid.

The formula for the volume of a pyramid of *any shape* is:

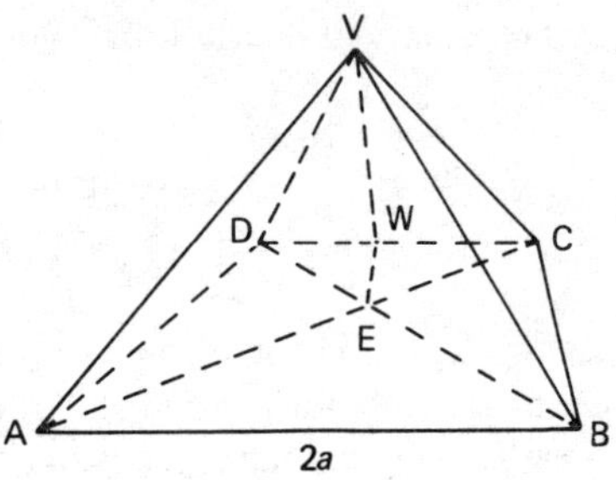

Fig. 59a

$\frac{1}{3} \times$ area of base $\times$ perpendicular height.

In our case this reduces to $\frac{1}{3} \times \text{AB}^2 \times \text{EV}$. . . . . (1)

To make it easier to follow, Figs. 59b and 59c, which form part of Fig. 59a, have been extracted and are shown alongside.

We see immediately that $\text{AB}^2 = (2a)^2 = 4a^2$. . . . . (2)

To find EV takes a little longer.

In $\triangle$ABC $\quad \text{AC}^2 = \text{AB}^2 + \text{BC}^2 = (2a)^2 + (2a)^2 = 4a^2 + 4a^2$
$= 8a^2$ (Pythagoras' theorem)

$\therefore \quad \text{AC} = \sqrt{8}a = \sqrt{4 \times 2}a = 2\sqrt{2}a,$

and hence $\quad \text{AE} = \frac{1}{2}\text{AC} = \sqrt{2}a$ (diagonals of a square)

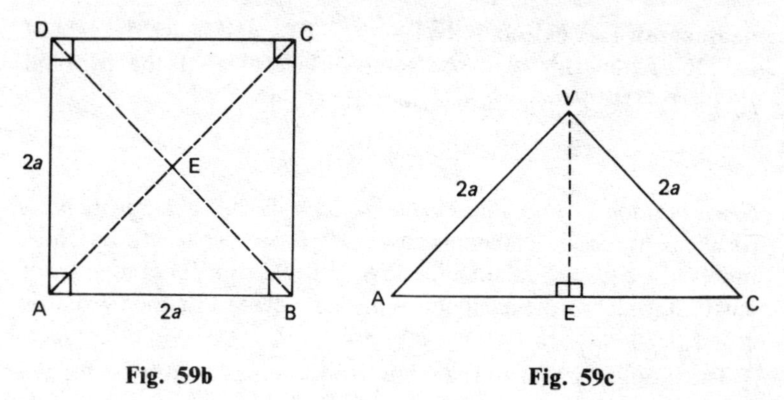

Fig. 59b **Fig. 59c**

Now consider $\triangle$AEV, in which AE $= \sqrt{2}a$ and AV $= 2a$ (given)

$$\text{AE}^2 + \text{EV}^2 = \text{AV}^2 \qquad \text{(Pythagoras' theorem)}$$

i.e. $(\sqrt{2}a)^2 + \text{EV}^2 = (2a)^2 \Rightarrow 2a^2 + \text{EV}^2 = 4a^2$

$\therefore$ $\text{EV}^2 = 4a^2 - 2a^2 = 2a^2$

(on taking $2a^2$ from each side)

so $\text{EV} = \sqrt{2}a$ . . . .(3)

Substituting from (2) and (3) in (1), we have Volume of VABCD

$$= \tfrac{1}{3} \times 4a^2 \times \sqrt{2}a$$

$$= \frac{4\sqrt{2}}{3}a^3 \text{ (cubic units)}$$

Fig. 59d (different scale)

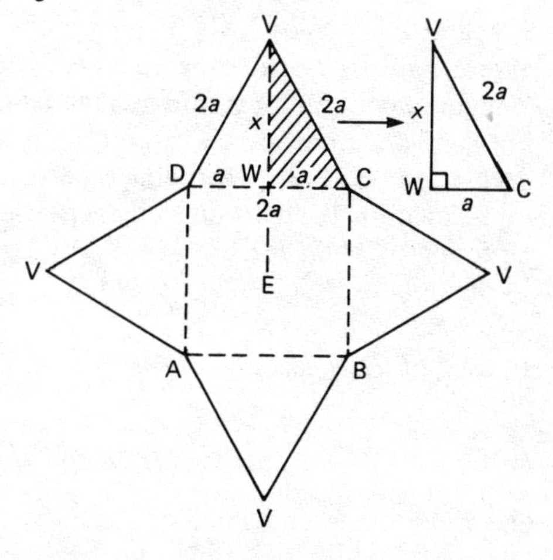

Problem Fig. 59d shows the net, which consists of four equilateral triangles of side $2a$ around a square, also of side $2a$. By drawing VW (which is perpendicular to DC) and observing that W is the midpoint of DC, calculate the length of VW, and hence the area of $\triangle$VDC. From this, find the total surface area of the pyramid ($\sqrt{3} \simeq 1.7321$). (It is approximately $10.93a^2$.)

Generalisation for the volume of any regular pyramid on a square base. As above, the base is taken as having edges each of length $2a$. Now, however, we take the slant edges VA, VB, VC and VD each to be of length b, which is independent of a. The volume V is then given by $V = \frac{4}{3}a^2\sqrt{b^2 - 2a^2}$.

The proof is similar to that above and is left as an exercise for the reader.

Question If F_n, V_n and E_n are the numbers of faces, vertices and edges, respectively, in accordance with the lettering of the question on page 122, what is the value of $F_n + V_n - E_n$ for the square pyramid above? (N.B. V_n and E_n used here are not related to letters V and E in Fig. 59a, which is the one needed to answer this question.)

Euler's Theorem Leonhard Euler (1707–82) was an eminent Swiss mathematician, who also studied astronomy and optics, *inter alia*.

Using the same notation as above, Euler's theorem states that in any simple polyhedron,

$$F_n + V_n - E_n = 2.$$

It will now be realised why the questions on $F_n + V_n - E_n$ have been asked – namely, to enable the reader to notice for himself that the value 2 might be constant. It is true, as a matter of fact, for many polyhedra but there are three notable exceptions, namely: (1) the small stellated dodecahedron, (2) the great dodecahedron, and (3) the great stellated dodecahedron. (Full details of these polyhedra, and many others, will be found in *Mathematical Models*, by H. Martyn Cundy and A. P. Rollett, Oxford University Press.)

In each of these cases,

$$F_n = 12,\ V_n = 12 \text{ and } E_n = 30$$

$$\therefore \quad F_n + V_n - E_n = 12 + 12 - 30 = -6,$$

and Euler's theorem is not satisfied.

EXERCISE 2

1 A cubical box is to be made out of plywood. Each edge of the box is to be 1 ft 9 in long. If a lid is to be included, find the area of plywood needed, in square feet. The faces are cut from a sheet measuring 6 ft by 4 ft. What percentage is wasted?

2 A lidless cuboidal wooden box is made of planks 10 cm wide. The length, breadth and height of the box are 0.9 m, 0.5 m and 0.3 m respectively. Neglecting the thickness of the planks, find
  (i) the area of wood needed, in square metres,
  (ii) the total length of planking needed, in metres,
  (iii) the cost of the planks at 65 p a metre run (i.e. for each metre in length),
  (iv) the length of a main diagonal of the box (correct to two decimal places).

3 Using a net similar to that in Fig. 57b, construct a cube, of edge 4 cm, from thick paper or thin card. Score along the dotted lines to make it easier to fold. Do not yet glue the cube, but hold it in shape by means of elastic bands or some other device. Mark a point X on one face and another point Y on a different face. Get someone to hold a piece of thread along the shortest and tightest possible path from X to Y, on the outside of the box. Now pencil a line alongside the thread. Remove the thread and elastic bands, etc., and lay the net out flat again. What do you observe about the pencil line from X to Y? (The path is called a *geodesic*. It is the shortest distance between two points on a surface, whether or not the surface is curved or angular.)

Note: If you intend to keep your cube afterwards, do not forget to add tabs for gluing. If they are tucked in, they do not affect experiments like the above.

4 The diagram overleaf, which is not to scale, represents a net of a church steeple, with vertex V and square base ABCD; AB = 10 m and AV = 13 m.

  (i) Calculate the total slant area (i.e. the four triangles) of the steeple. This would be important in estimating the cost of, say, re-slating or re-tiling.
  (ii) Calculate the height of V above ABCD.

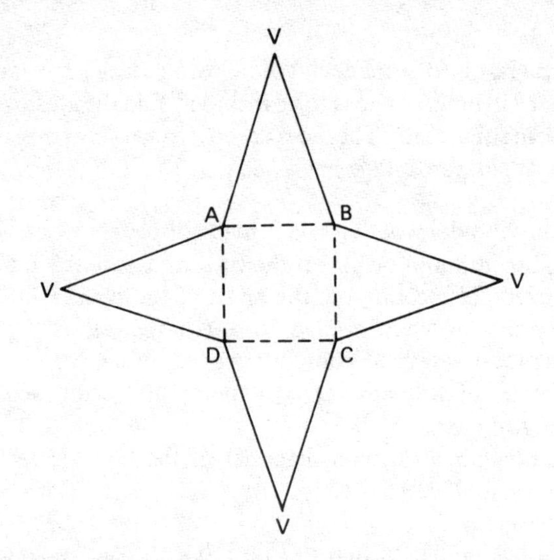

Definition A *right prism* is a solid which has a uniform cross-section, the end sections being perpendicular to the longitudinal edges, which form rectangles. Apart from glass prisms, which have triangular end-sections, other common examples are greenhouses (usually) and girders. The volume of a right prism is area of an end section × length of prism (Fig. 60).

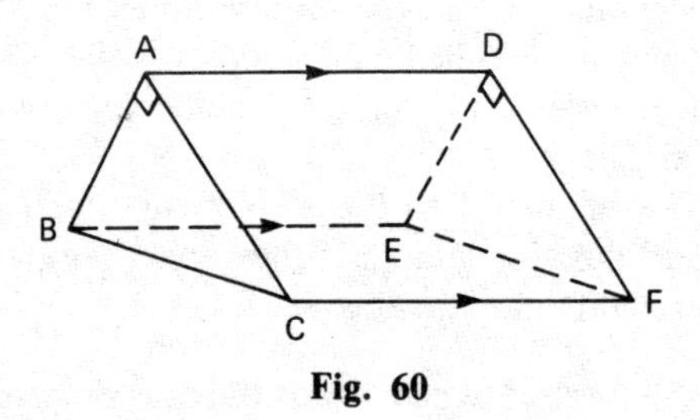

Fig. 60

5 Calculate the volume of the prism illustrated in Fig. 60, given that angle A = 90°, AB = 9 cm, AC = 12 cm and AD = 25 cm. Give the answer in cubic centimetres.

Find the total surface area and also draw a net of the prism to a linear scale of $\frac{1}{4}$ actual size. Hence make a model of the prism.

Definition A *tetrahedron* is a solid with four triangular faces, i.e. it is a triangular pyramid. If all the faces are equilateral triangles it is a *regular tetrahedron* (Fig. 61).

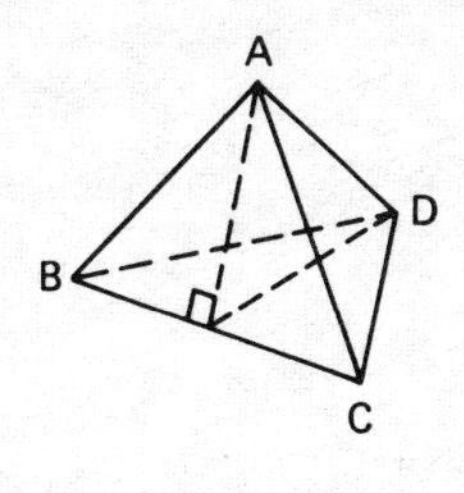

Fig. 61

6 A regular tetrahedron ABCD has AB = $2a$. Show that the total surface area is $4\sqrt{3}a^2$ sq. units. (A constructional hint is given in Fig. 61.)

Draw a net for the tetrahedron, choosing any convenient scale, and try the geodesic experiment.

*The volume of this tetrahedron is $\frac{2\sqrt{2}}{3}a^3$. The proof is a little more searching.

7 Do the prism and the tetrahedron, in questions 5 and 6 above, satisfy Euler's theorem $F_n + V_n - E_n = 2$?

8

The Electronic Calculator and its Use

1 An historical introduction to the electronic computer

Simple calculating machines existed thousands of years ago. They were mainly variants on the abacus (a counting frame) and they have continued to be employed in various parts of the world until the present day. Now, however, the abacus is used mainly by the Chinese. Moving nearer home, it has only been within the last fifty years that the widespread practice of giving young children abaci, to aid them in numeracy, has disappeared; it is a pity that the custom did not survive!

Fig. 62 shows part of a complete archaic Roman bronze abacus. The lower part of the counting frame has *four* buttons in each vertical slit; starting from the *left*, each button represents 1, 10, 100 . . . units. The upper part of the frame has *one* button in each vertical slit, representing, from left to right, 5, 50, 500 . . . Buttons at the bottom of a slit do not count; those at the top score.

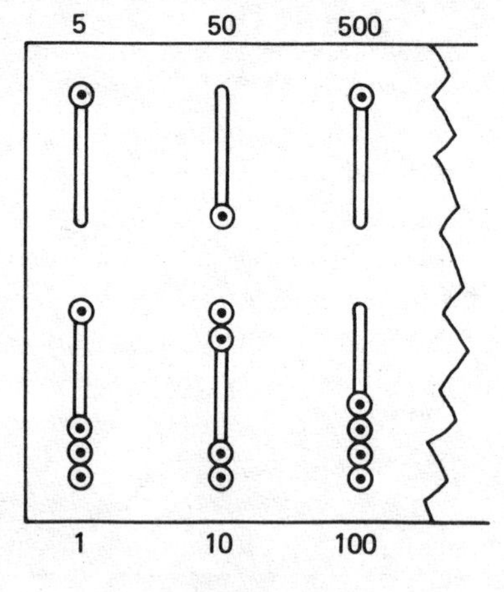

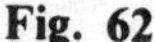

Fig. 62

The score shown is:

Col. I: 5 + 1 = 6; Col. II: 0 + 20 = 20;
Col. III: 500 + 0 = 500; TOTAL 526.

The reader who is puzzled as to why there are only four buttons, and not nine, in the lower slits, is reminded that the Romans counted in fives, not in tens; they had special symbols V, L and D for 5, 50, 500, respectively, as well as X, C for 10, 100, etc.

Various attempts were made to develop more sophisticated calculators through the ages, one of the best-known efforts being that of Charles Babbage (1792–1871), an Englishman, who tried to perfect a mechanical computer working on a system of toothed gears. He hoped to solve difficult mathematical problems in this way. Having devoted 36 years to this end, during which time he spent a small fortune, he failed. His most interesting experimental calculator has, however, been preserved.

In the year 1847, George Boole (1815–64), a self-taught British mathematician, produced a work entitled *The Mathematical Analysis of Logic*. It remained almost neglected for nearly a century, until in 1936 the theory of the electronic computer was developed at Manchester University. In consequence, *Boolean algebra*, as it is called, at last gained recognition as to its true value, for its logical presentation was exactly what was required for a computer of this type. At the Cavendish Laboratories in Cambridge an experimental computer utilising thermionic valves was built. The first *full-scale* computer is said to have been the ENIAC (Electronic Numerical Integrator and Calculator) constructed in 1946 at the University of Pennsylvania.

Initially computers were large and cumbersome affairs housing hundreds – sometimes thousands – of thermionic valves. It required the invention of the transistor (a solid state amplifying electronic device) by Bardeen, Brattain and Shockley, of the USA, followed later by the miniaturisation of micro-circuitry, to reduce computers to moderate dimensions, and to facilitate the development of the present pocket-sized electronic calculators.

The difference between a computer and a calculator is worthy of mention. A computer may be programmed (i.e. preset with a *program* – the internationally agreed spelling – to carry out a series of operations on each of a long series of different sets of data, whether the data be mathematical, scientific or commercial). A calculator requires the insertion of instructions suitably interspersed with data *each time* it is given information.

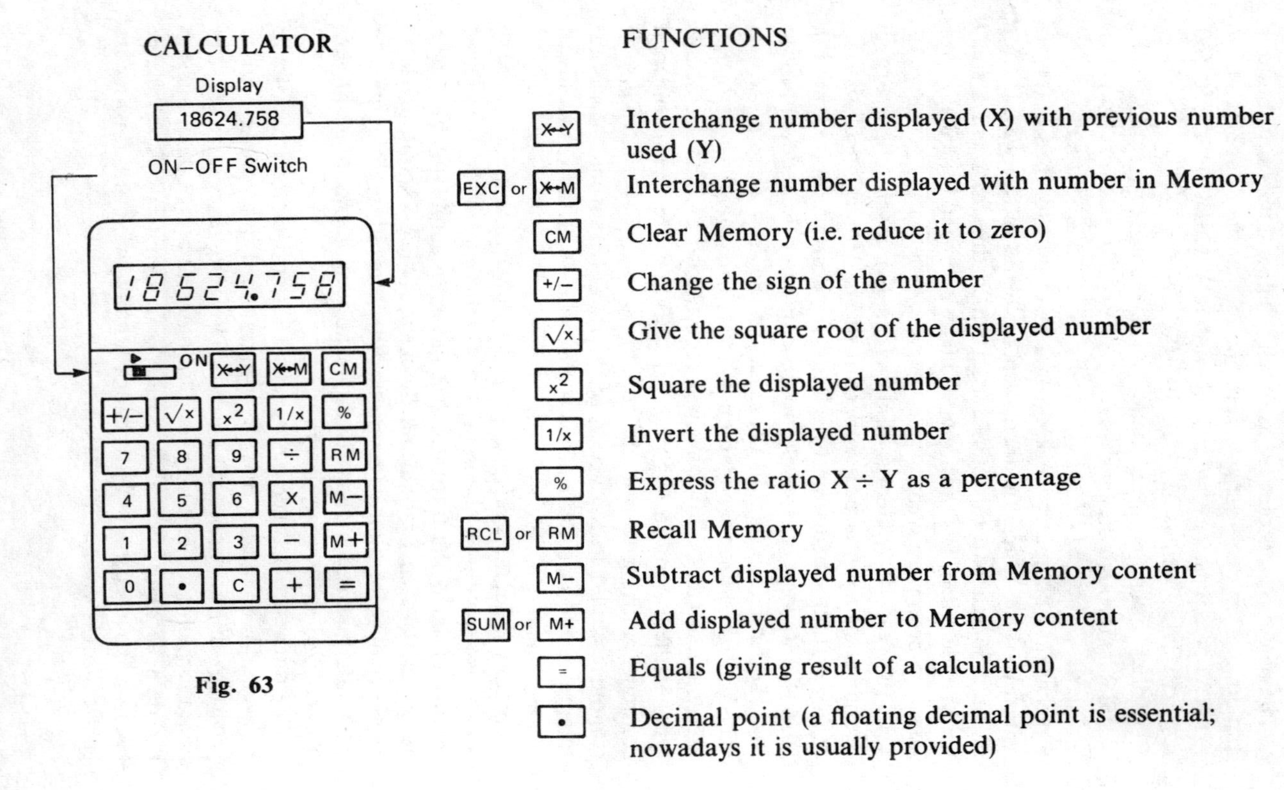
CALCULATOR
Display
18624.758
ON–OFF Switch
18624.758
ON
X↔Y
X↔M
CM
+/–
√x
x²
1/x
%
7
8
9
÷
RM
4
5
6
X
M–
1
2
3
–
M+
0
•
C
+
=
FUNCTIONS
X↔Y Interchange number displayed (X) with previous number used (Y)
EXC or X↔M Interchange number displayed with number in Memory
CM Clear Memory (i.e. reduce it to zero)
+/– Change the sign of the number
√x Give the square root of the displayed number
x² Square the displayed number
1/x Invert the displayed number
% Express the ratio X ÷ Y as a percentage
RCL or RM Recall Memory
M– Subtract displayed number from Memory content
SUM or M+ Add displayed number to Memory content
= Equals (giving result of a calculation)
• Decimal point (a floating decimal point is essential; nowadays it is usually provided)

Fig. 63

2 The use of an electronic calculator

These calculators rapidly handle heavy arithmetic, but they vary considerably in the number of functions made available to the user. We shall firstly examine a machine having a good quota of operational keys, but lacking logarithmic and trigonometric functions, which will be considered later. It would enable the user to solve many everyday problems (Fig. 63).

In addition to the above thirteen functions, there are five others. The first four, reading down part of column 4 of our hypothetical keyboard, are:

[÷] divide
[×] multiply (this must not be confused with *functions* containing the letter X or x – lower-case letters being used when space on the key is limited)
[−] subtract
[+] add

The final function needs special mention:

[C] (sometimes shown as [C/CE]) means Clear *or* Clear last Entry. It applies to direct access, *not* to memory, which can only be cleared by the [CM] key mentioned earlier. The use of the [C] key will be shown in various examples and summarised then.

The other ten keys are merely the numbers 0, 1, 2 . . . 9. All numbers on the display are shown as integers or in decimal form, e.g. 2704 or 31.68 or .33333333 (for $\frac{1}{3}$ in an eight-digit display). The decimal point always appears to be at the bottom of the line, not in the middle, in electronic calculators.

The reader should work all the following examples as he goes along.

Example 1 Calculate the values of the following expressions and, where necessary, give the results to two decimal places only. (The keys are pressed in the order shown.)

(i) 47×3.1

4	7	×	3	.	1	=

145.7

(ii) $17.28 + 219.062 - 104.256$

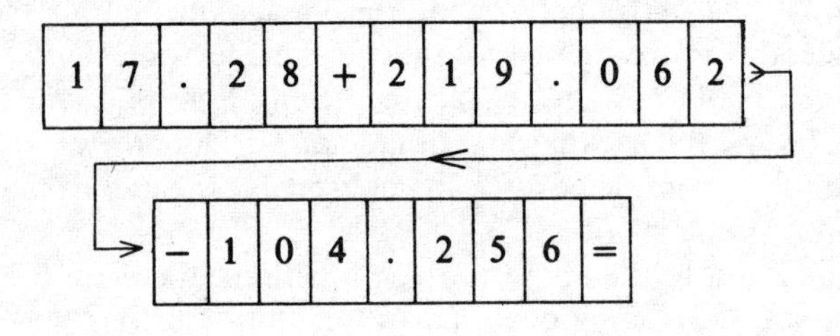

132.09

(iii) $5.08 \div 14 + 4.38$

5	.	0	8	÷	1	4	+	4	.	3	8	=

$4.7428571 \simeq \mathbf{4.74}$

(iv) $\dfrac{87 - 42.93}{36.4}$

8	7	−	4	2	.	9	3	=	÷	3	6	.	4	=

$1.2107143 \simeq \mathbf{1.21}$

(v) $\dfrac{1}{19} - \dfrac{1}{7}$

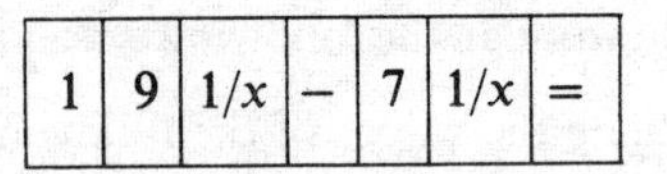

$-0.0902256 \simeq \mathbf{-0.09}$

The explanation is that the functions are applied *after* the relevant numbers and that they operate immediately, in the cases of [$\sqrt{x}$], [x^2], [$\frac{1}{x}$] (which is shown on the keyboard as [1/x]). It is not necessary to press [=] to obtain the reciprocal ($1/x$), but we must do so to finish the subtraction.

(vi) $(0.729)^2$

.	7	2	9	x^2	=

$$= 0.531441 \simeq \mathbf{0.53}$$

When there is a *zero only* in front of the decimal point, we do not need to depress it, as it will automatically appear on the display. Likewise when only *one* function of this type is used, the = sign does not need to be depressed.

(vii) $\sqrt{62.98} - \sqrt{47.92}$

6	2	.	9	8	$\sqrt{x}$	−	4	7	.	9	2	$\sqrt{x}$	=

$$1.0135666 \simeq \mathbf{1.01}$$

(viii) $\sqrt{3780.93 \times 0.0175}$

3	7	8	0	.	9	3	×	.	0	1	7	5	=	$\sqrt{x}$

$$= 8.1342655 \simeq \mathbf{8.134}$$

In this case we must complete the multiplication by depressing the = sign, before we take the square root (which instantly gives the answer).

(ix) $\frac{1}{2}(47.4 + 28.6) \times 15$ (the area of a trapezium, page 115)

4	7	.	4	+	2	8	.	6	=	×	1	5	÷	2	=

We must evaluate the bracket first. In a few calculators it is not necessary to put [=] immediately after $47.4 + 28.6$, as the multiplication sign takes care of the addition, but in the interests of safety it is wise to insert the = sign as shown.

(x) $\sqrt{a^2 + b^2}$, where $a = 4.736$ and $b = 5.927$. This formula will give the length of the hypotenuse of a right-angled triangle, when the lengths of the other sides are known.

$$\sqrt{(4.736)^2 + (5.927)^2}$$

4	.	7	3	6	x^2	+	5	.	9	2	7	x^2	=	$\sqrt{x}$

$$= 7.5867664 \simeq \mathbf{7.59}$$

(xi) Reduce $1 + \frac{r}{100}$ to decimal form, when $r = 4\frac{3}{8}$ (Useful for compound interest – see later).

3	÷	8	+	4	=	÷	1	0	0	+	1	=

1.04375 (all decimal places needed)

Repeated calculations using a constant function and number can be rapidly undertaken, although methods may vary substantially from one type of instrument to another. Nevertheless, the following seems to be common to numerous calculators. We need an extra key [K] (constant key), or some similar title. This key is *not* illustrated in Fig. 63.

(xii) Suppose we carry out four simple operations on a sequence of numbers, say, 1.2, 4.4, 7.8 . . . . To them, in turn, we shall add 1.6; then subtract 1.6 from each; then divide each by 1.6; and then (to complete the quartet) multiply each by 1.6. These illustrate the application of the *four rules* (Chapter 2).

We proceed as follows:

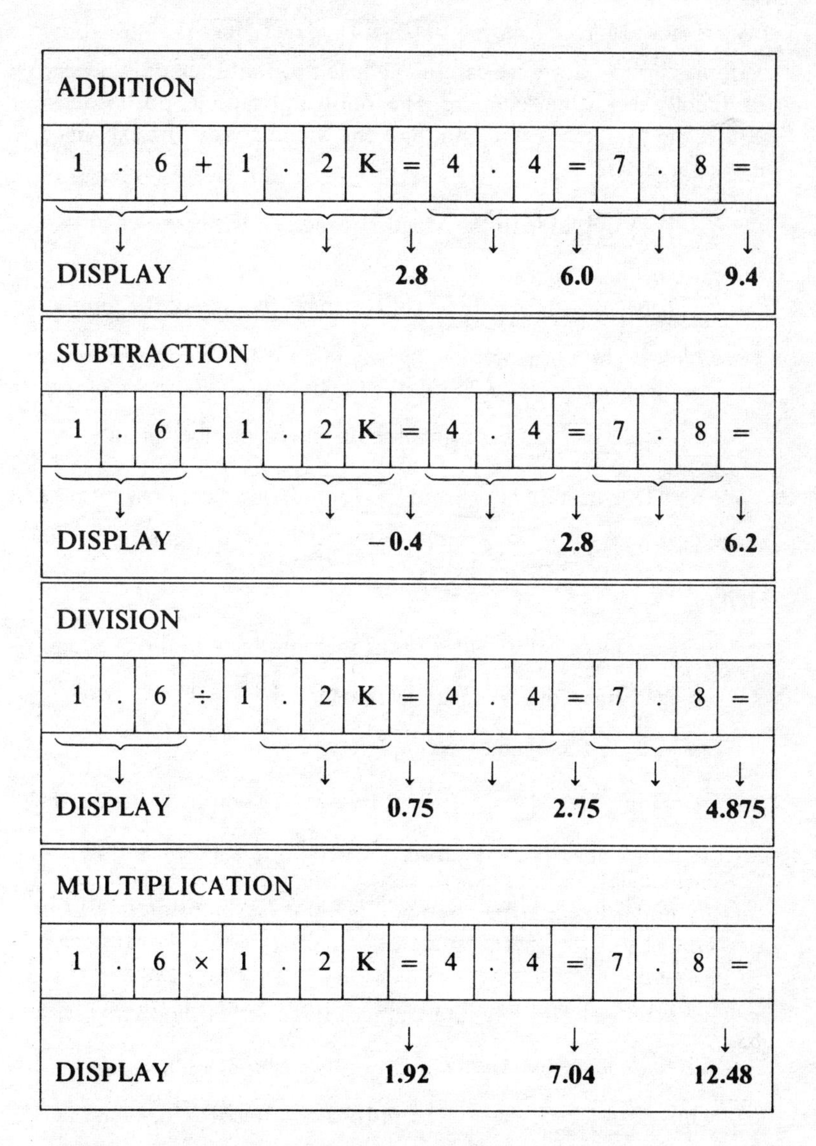

Slight variation in the names *and uses* of memory keys may be found, but the following seems to be fairly standard procedure.

The reader will probably have observed that we have not as yet used any of the Memory keys [M+], [M−], [CM], [EXC]. The above examples, (i) to (xii), were carefully chosen so that they only needed the directly accessible keyboard. The following examples do, however, require the Memory. Firstly, though, the Memory keys are explained in greater detail.

[M+] or [SUM] adds to the Memory exactly what is shown at that time in the display.

[M−] does the same as [M+], but changes the sign of the number which is shown in the display (e.g. 61 on the display becomes −61 in Memory; −3.25 on display becomes 3.25 in Memory).

[M+] and [M−] do *not* affect the displayed number in any way.

We may feed as many [M+] and [M−] numbers as required into the Memory; thus, 17 [M+] 2.64 [M+] 3.82 [M−] gives a total of 15.82 in the Memory.

[RCL] or [RM] recalls what is in the memory back to direct access, whilst leaving this quantity in the Memory itself.

[CM] clears the Memory completely.

Note: Nowadays [M−] often seems to be omitted from the keyboard and, in such a case, insertion of [−] [2] [SUM] puts 2 (*not* −2) in the Memory! There is, however, a subterfuge which proves to be satisfactory: [2] [+/−] [SUM].

[EXC] *or* [x↔M] This key interchanges the number *on display* with the *contents* of the *Memory*. Press once, and [X] becomes [M] whilst [M] becomes [X]; press again and the reverse takes place (an as-you-were situation). Some calculators do not have this key, but see the note to Example 2(ii), below.

One of the great advantages of the Memory is that its contents sit there peacefully, no matter what is happening in the direct access, unless one presses a Memory key.

Example 2 Evaluate the following expressions, using Memory keys when these are necessary or helpful. Give the answers correct to three places of decimals, where relevant. (See also the alternative easier method available on many calculators: Example 2 repeated below.)

(i) $43.92 \times 2.83 + 65.8 \times 9.07$

In accordance with the BODMAS rule, we must carry out multiplication before addition. Thus, we work out 43.92×2.83, transfer it to memory [x→M] and, as Memory has nothing in it at this stage, direct access becomes zero; then we calculate 65.8×9.07, and add back the Memory using + and RM keys, in that order.

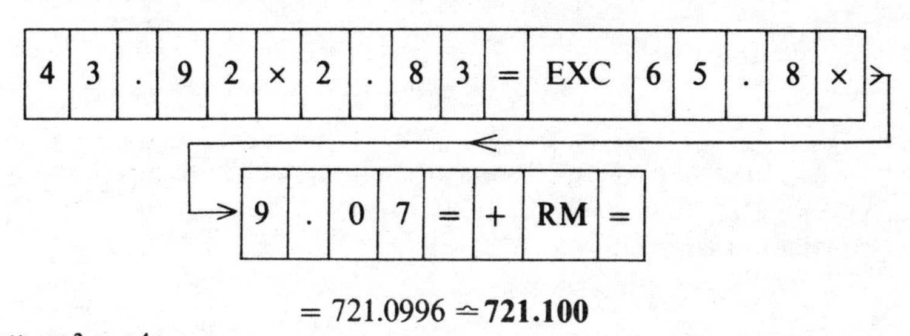

$= 721.0996 \simeq \mathbf{721.100}$

(ii) $(5\frac{2}{7} - 3\frac{4}{9}) \div 6$

We evaluate the contents of the brackets and then divide by 6. Note, however, that if we put $5 + 2 \div 7$, some calculators may do the addition first, i.e. $5 + 2(= 7) \div 7 = 1$, which gives $5\frac{2}{7} = 1$! We proceed thus $2 \div 7 + 5$, and all is well for any normal machine.

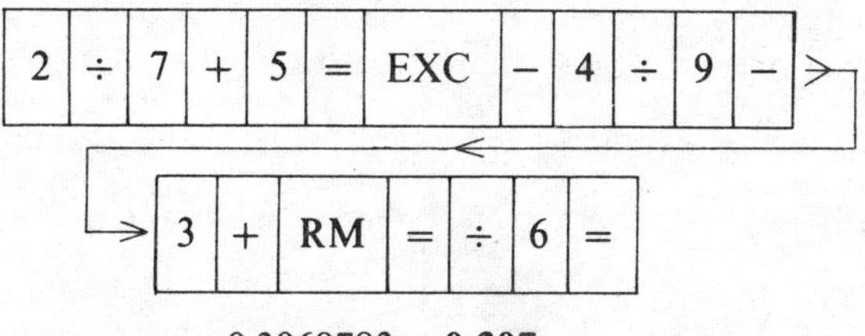

$0.3068783 \simeq \mathbf{0.307}$

Note: Quite a lot of calculators do not include the EXC key, which is a pity. Nevertheless, in many cases, i.e. *when merely wishing to transfer direct access to memory and at the same time clearing memory ready for a further step*, we can use the key M+ or SUM instead. It is applied at the same place as EXC and need not affect the number of keys used. (See (iii) below.)

(iii) $36 \times 45 + 72 \times 37 + 44 \times 39$

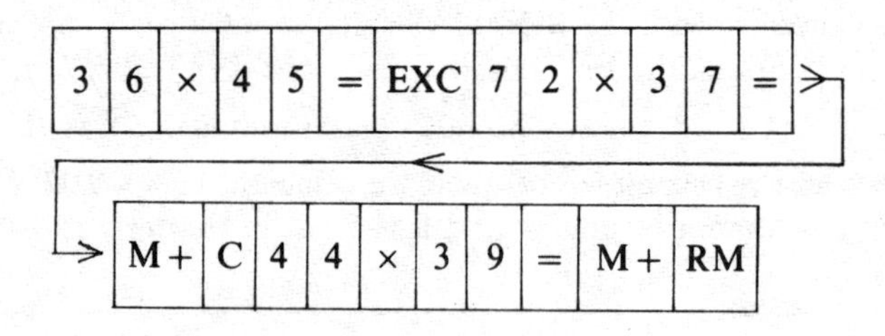

= **6000**

The principal point about this example is that EXC can only be used for the first multiplication, when the memory is zero. The explanation is as follows:

		Access	Memory
Stage 1	36 × 45 =	1620	0
	EXC	0	1620
Stage 2	72 × 37 =	2664	1620
(If we now used EXC we would have 1620 2664 which is useless, as we have reversed the entries)			
Hence	M+	2664	4284
(Clear Access)	C	0	4284
Stage 3	44 × 39 =	1716	4284
	M+	1716	6000
	RM Display	6000	6000

At this point it is worth mentioning that *many modern electronic calculators do not require the use of a Memory for Example* 2 (*i*), (*ii*) *and* (*iii*), although the examples can be done in the ways just shown. The rather shorter method available on these instruments is now given for the same three examples.

Example 2 *Repeated* for the alternative method, when available on the calculator.

(i)	4	3	.	9	2	×	2	.	8	3	+	6	5	.	8	×	9	.	0	7	=		**721.1**
(ii)	2	÷	7	+	5	−	4	÷	9	−	3	=	÷	6	=	→							**0.307**
(iii)	3	6	×	4	5	+	7	2	×	3	7	+	4	4	×	3	9	=	→				**6000**

The following Example 3(i) and (ii) do not lend themselves to the shorter alternative method, nor will all simple machines solve them without writing down part results. Many will, however, do so as follows.

Example 3 Evaluate the following expressions, using Memory keys, to three decimal places.

(i) $\dfrac{52 \times 19 - 36 \times 29}{16 \times 58 + 23 \times 73}$

(ii) $\sqrt{\dfrac{(a+b)(a-c)}{bc}}$, where $a = 7.41$, $b = 3.98$ and $c = 5.264$.

(i)	5	2	×	1	9	−	3	6	×	2	9	=	EXC	1	6 →

→ ×	5	8	+	2	3	×	7	3	=	÷	RM	=	1/*x*

−0.021 (correct to three decimal places)

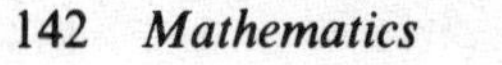

(ii) $$\sqrt{\frac{(a+b)(a-c)}{bc}} = \sqrt{\frac{(7.41+3.98)(7.41-5.264)}{3.98 \times 5.264}}$$

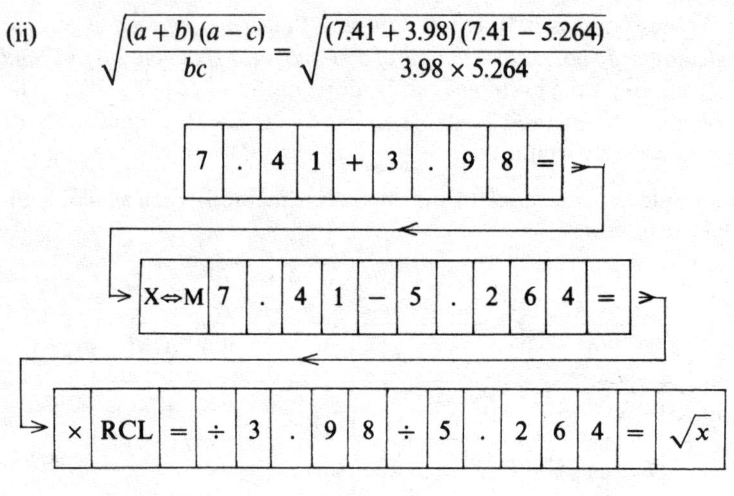

1.080133 ≏ **1.080** (correct to three decimal places)

Note alternative keys X ⇔ M and RCL, used instead of EXC and RM, respectively.

3 Important notes on the use of a calculator

1 Although it is possible to buy a very cheap calculator, provided with only a few keys, the reader who wishes to use his instrument for a variety of personal, domestic, financial or business problems, will soon get frustrated with such limited facilities. The composite keyboard illustrated earlier will, however, handle most simple, and many reasonably elaborate, arithmetical calculations. The *exact* pattern will not be found on the market but, as long as all or nearly all of the recommended keys are present, this does not matter. (The Exchange – or interchange – and Memory keys may vary slightly in name and usage.) There may be a few extra keys as well. If these should happen to be common logarithms (to the base 10) and trigonometric functions, they could be helpful later on. For the average user, however, I do *not* feel that instruments which are indicated as *programmable* are necessary. They are more costly. Furthermore, the programs available for a pocket-sized calculator can be tricky to operate and strictly limited in use. This is hardly surprising – full scale computers are expensive and appreciably larger; they also require skill.

2 Depressing the [C] or [CE] key once will clear *the last entry* in direct access. To clear *all* the contents of direct access, depress the above key *twice*. This key does not affect the contents of the memory.

3 To clear memory, depress [CM] if present, or switch off.

4 Always switch off after use, unless further work is to follow after a *very* short time, as in (2) above. Switching off clears everything; it also safeguards the calculator and the battery.

5 Before doing the exercises below, the reader who has had little experience of calculators is advised to study each of the above *worked* examples in detail. He should then try to do each one himself, referring back to the relevant layout if he gets into difficulty. It takes a certain amount of practice to use a calculator efficiently, but mastering of the technique is not boring. A problem may frequently be resolved by various combinations of keys, some taking longer than others. One can get quite a lot of satisfaction in finding a neat solution.

6 If a calculator is to be put away for a long period of time, say a month or more, *it is advisable to remove the battery* (or batteries) so that damage cannot arise within the instrument, as a result of chemical reaction caused by cell deterioration.

EXERCISE 1

In questions 1 to 21 inclusive, use direct access (but not memory) to evaluate the expressions, giving the result (i) to the accuracy of the calculator, (ii) correct to two decimal places, if there are more than this number on display:

1 $372+48$ 2 $64+1938+269$

3 $425-286$ 4 $63+48-17$

5 $249-76-180$ (Note that a minus sign appears to the left)

6 $27.3+580.26+0.779$ 7 $647.8-229.13-418.67$

8 384×76 9 21.09×4.67

10 $35.28\times 4.55\times 9.6$ 11 4056×0.005

12 $832.74\div 37$ 13 $6.95\div 207-0.032$

14 $\dfrac{36-47.36}{7.1}$ 15 $\dfrac{14.65+83.11}{26\times 4.7}$

16 $1-\frac{1}{3}+\frac{1}{5}-\frac{1}{7}+\frac{1}{9}$

17 $46\left(\frac{1}{27}-\frac{1}{53}\right)$ (Firstly calculate the brackets)

18 $\sqrt{p^2-q^2}$ when $p = 8.7$ and $q = 5.3$

19 $\sqrt{9^2+(4.5)^2+3^2}$ (the diagonal of a brick, in inches)

20 $\sqrt{0.007283}-\sqrt{0.005648}$

21 $\sqrt{473.98\times 0.00677}$

For questions 22–32, use the Memory where necessary. Give answers correct to three decimal places, unless the answers are integers:

22 $61\times 23+8\times 712$

23 $36.4\times 2.8-24.9\times 3.6$

24 $(61.93-73.8)\div(7.23+8.92)$

25 $7\frac{3}{11}-5\frac{4}{9}$

26 $18\frac{5}{7}\times 3\frac{4}{19}$

27 $\frac{7}{24}\div\frac{11}{13}$

28 $\frac{2.074}{81.56-69.36}$

29 $\left\{\frac{16.1}{23}+\frac{38.7}{116.1}\right\}\div(0.924-1.646)$

30 $3(4.093)^2+5(2.867)^2$

31 $\sqrt{3a^2-5b^2}$, when $a = 6.82$, $b = 4.97$

32 $\sqrt{k^2/lg}$, when $k = 4.6$, $l = 7.3$, $g = 9.8$

4 Some special problems in using the calculator

Correction of a numerical error in entry

Example 4 Suppose we have a simple expression such as $37+46-53$ and in error we put $37+46-43$. Having noticed in the display that the 43 was wrong, *and provided that we have not put any more entries in the machine*, we can easily correct the slip:

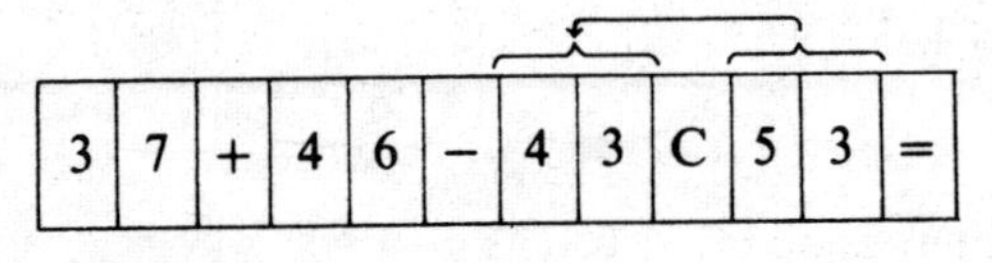

The [C] cancels the previous number, 43, only. We then immediately put 53 and finish the calculation with [=]. If, however, another function has been entered before the error is spotted, one must start again from the beginning. (*Now turn to Exercise 2, Part 1.*)

Overflow

This tends to occur when very large numbers are employed. It is probably best explained by an illustration. High numbers in mathematics and in science are often represented as a number with one *digit only* before the decimal point, this number being multiplied by a power of 10. One such number is the speed of light; this is 299 800 kilometres per second (km/s). If we wish to express this in metres per second we get $299\,800 \times 1000 = 299\,800\,000$ m/s. Now nine figures are required to express this in full. Most calculators only display up to eight figures. Suppose, therefore, on our calculator we put 299 800 multiplied by 1000. If our machine has a good capacity for information we may expect to get, dependent upon its make, either

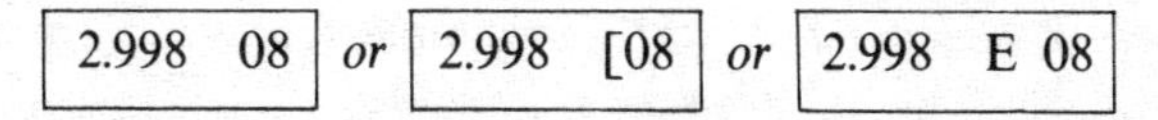

Each of these expressions means the same thing, namely 2.998×10^8 (which is, of course, the same as 299 800 000). The method of presentation is called *standard*, or *scientific, form.*

Standard form is normally used to express very large or very small numbers in an easily understandable way, which avoids the use of a long row of zeros. It consists of two parts multiplied together: the first part is a number in decimal form, having a single figure (*not* zero) in front of the decimal point, followed by as many decimal places as are required; the second part consists of a power of 10, i.e. 10^n, where n is a positive or negative *integer*.

Example 5 Find the approximate value of $987\,654 \times 456\,789$, correct to four decimal places. This gives

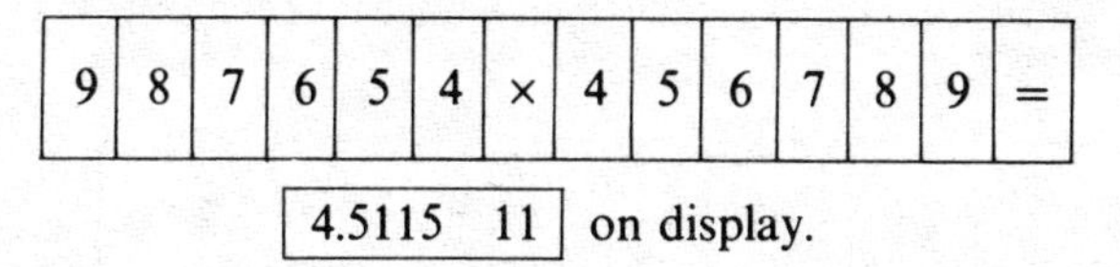

$\therefore$ the value required is $\mathbf{4.5115 \times 10^{11}}$ correct to four decimal places.

Different calculators vary substantially with regard to what is given on display when, in the case of overflow, we press the [=] key. If the display varies significantly from the above style, the reader will need to study the manufacturer's instructions carefully.

Example 5 illustrates a case in which n is positive. An example of a large negative power of n is given by the conversion of an electron-volt (the work done on an electron when it passes through a rise, in potential, of 1 volt) to the corresponding SI unit.

$$1\ \text{ev} = 1.60219 \times 10^{-19}\ \text{J}$$

(ev = electron-volt; J = joule, the unit of work, formerly 10^7 ergs).

The electron-volt occurs in nuclear physics, but we shall not discuss it further. (*Now turn to Exercise 2, Part 2.*)

Calculations involving powers

We shall, in the next chapter, apply the use of our calculator to practical problems which involve more elaborate mensuration. A start will, however, be made here with an example, followed by numerical illustrations of the methods of usage of the [×] and [x^2] keys, when powers higher than 2 are needed; the procedure is, however, much simpler if the calculator has an [x^y] key (see page 148).

Example 6 The volume of a sphere is given by the formula $V = \frac{4}{3}\pi r^3$. If $\pi = 3.14159$ and $r = 3.54$ cm, find the volume, giving the result correct to one decimal place. (The object used was an old cricket ball found at the back of a kitchen drawer; it was not truly spherical!)

We firstly find $r^3 = (3.54)^3$. This can be done as follows:

$r \times r^2$, i.e.

3	.	5	4	×	x^2	×

44.361 865 (using × and x^2)

Working the whole thing:

3	.	5	4	×	x^2	×	3	.	1	4	1	5	9	×	4	÷	3	=

$$185.82238 \simeq \mathbf{185.8\,cm^3}$$

Although on many calculators one can alter the order of the operations above, it is not universally true. Hence it is best to set the power operation in the calculator initially.

Incidentally, when using eight digits in different calculators there may be a small difference in the last digit; e.g. 185.82238 may appear as 185.82237. Also, many calculators have a key for π (see page 156).

Repetitive use of [×] *and* [x^2]: consider 7^7, 7^8 and 7^9, for example:

(i) 7^7 We have

7	x^2	x^2	x^2	÷	7	=

823 543

(ii) 7^8 We have

7	x^2	x^2	x^2

5 764 801

(iii) 7^9 We have

7	x^2	x^2	x^2	×	7	=

40 353 607

Display

The explanation of the abbreviated steps follows immediately from the laws of indices:

(i) $\{(7^2)^2\}^2 \div 7 = \{7^{2\times 2}\}^2 \div 7 = \{7^4\}^2 \div 7 = 7^8 \div 7^1 = 7^{8-1} = 7^7$

(ii) $\{(7^2)^2\}^2 \quad = \{7^{2\times 2}\}^2 = \{7^4\}^2 = 7^{4\times 2} = 7^8$

(iii) $\{(7^2)^2\}^2 \times 7 = 7^8 \times 7^1 = 7^{8+1} = 7^9$

Alternatively, when we only possess calculators which do not have [x^2] or [x^y], we must use for, say, 7^7, the following clumsy process:

7	×	7	×	7	×	7	×	7	×	7	×	7	=

823 543

Example 7 Find the value of $2^{15} \times 3^3 \times 5^3 \times 11^3$.

The quickest way to break this down is probably in two parts:

$$2^{15} = 2^{16} \div 2 = [\{(2^2)^2\}^2]^2 \div 2$$

and $$3^3 \times 5^3 \times 11^3 = (3 \times 5 \times 11)^3 = (3 \times 5 \times 11) \times (3 \times 5 \times 11)^2$$

We then enter the first one and transfer to memory; enter the second one and recall the memory, finishing by multiplying in the usual way:

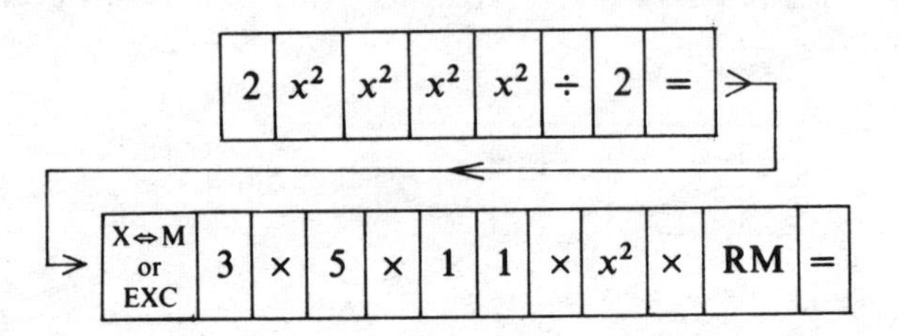

1.472 × 10^{11}, which is **147 200 000 000** (to four significant figures).

This is actually the number of cubic feet in one cubic mile. We could have done it more succinctly as 5280^3, but it would have been far less instructive than the prime factors used above. (Now he tells us!) The *exact* answer is 147 197 952 000. (*Now turn to Exercise 2, Part 3.*)

Use of $\boxed{x^y}$ *key* (sometimes called $\boxed{y^x}$, which makes no difference), when present.

This is very quick and easy.

Example 8 Find the value of $(7.839)^{11}$

We have

7	.	8	3	9	y^x	1	1	=

6.8686 09

$= \mathbf{6.8686 \times 10^9}$ (correct to five significant figures).

EXERCISE 2

Part 1 – Correction of Errors

The method is shown in Example 4 on page 144. It does not matter whether there are two, *or more than two*, previous numbers. These may include any or all of +, −, ×, ÷, etc., signs.

1 (a) On your calculator, enter 19×7. Whilst the 7 is on display, you notice that it should be 6. Without cancelling everything, correct the error and find the answer.

(b) Put into your calculator, $16 - 5 + 21$. Whilst the 21 is on display, correct it to 12 and complete the sum.

2 Enter $3.68 \div 0.151$. Before proceeding further, you observe that 0.151 should be 0.115. Correct this and complete the division.

3 Enter $7.93 - 15.86 - 6.238$. At this point you see that the last-named should have been 2.683. Correct this and find the result.

Part 2 – Overflow

4 Find the value of $7430 \times 2968 \times 395$. Give the result in standard form $A \times 10^n$ where A is a number (to be found) having one non-zero figure in front of the decimal point, and n is an integer also to be determined. Three places of decimals are required.

5 Express $(82\,593)^2$ in standard form.

6 Find the number of seconds in one year of 365 days. Now express this in standard form using as many decimal places as necessary.

How many seconds are there in four years, one of which is a leap year? Give the result in standard form.

7 Express the ratio of one ounce avoirdupois to one ton, giving the result in standard form. (16oz = 1 lb; 2240 lb = 1 ton)

8 The speed of light is approximately 186 326 miles a second. (i) How long does it take to go from Earth to the moon, taking its average distance from Earth to be 238 857 miles? (ii) How long does it take light to travel 1 mile, giving this result in standard form? (Both answers should be in seconds.)

Part 3 – Calculations involving Powers

N.B. The use of $\boxed{y^x}$ greatly simplifies the working here.

9 Find the values of (i) 5^4, (ii) 6^5, (iii) $(7.324)^3$, (iv) 12^6, (v) $(1.125)^8$.

10 Calculate, to the accuracy your instrument will give, (i) $(1.072)^{12}$, (ii) $(0.973)^{10}$, (iii) $(7.43)^3 \times (0.869)^4$.

11 Find the following, correct to four decimal places:

(i) $\left(\frac{3}{8}\right)^4$, (ii) $\left(\frac{2}{3}\right)^5$, (iii) $\left(\frac{11}{7}\right)^3 \times \left(\frac{8}{9}\right)^7$, (iv) $(3.215)^4 \times (0.0837)^4$.

12 Calculate the following, giving the results in standard form correct to three decimal places: (i) $17^4 \times (3.68)^6$, (ii) $4520 \times (1.1833)^9$.

In scientific work it may be just as important to express very small numbers in standard form as it is to record large ones in this way. When using a calculator for small numbers, however, we need to prepare our ground more carefully.

Example 9 Express the value of $(0.00012)^2 \times (0.7)^{-5}$ in standard form, correct to three decimal places.

This is easy if the $\boxed{y^x}$ key is present:

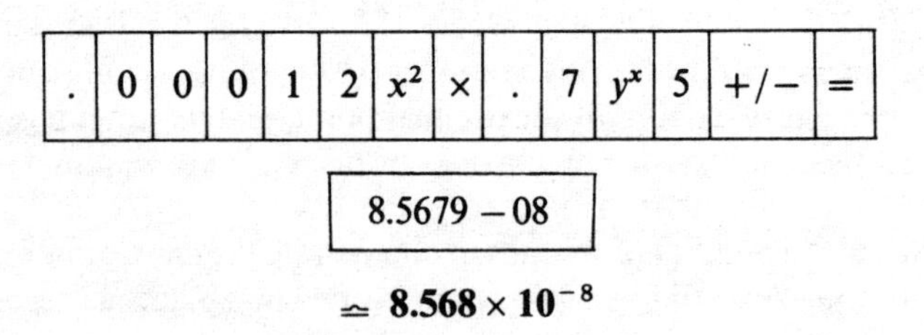

$$\simeq \mathbf{8.568 \times 10^{-8}}$$

If there is no $\boxed{y^x}$ key, we deal with the negative index firstly:

.	7	×	x^2	x^2	=	$1/x$	×	.	0	0	0	1	2	x^2	=

$\boxed{8.5679 - 08}$

giving the same result as before.

13 Find the values of the following, giving the answers in standard form, correct to 3 decimal places, where possible: (i) $(0.00728)^3$, (ii) $(0.00329)^5$, (iii) $(0.0045)^6 \times (0.0038)^{-2}$, (iv) $3^{-5} \times 4^{-3}$.

14 Find the value of

$$1+\frac{1}{2}+\frac{1}{3}+\ldots+\frac{1}{9}-2\sqrt{2}$$

correct to four decimal places. (*Hint*: make use of $\boxed{1/x}$ key.)

5 Supplementary observations

1 Keyboards of typewriters, other than specialised ones, are pretty well standardised, no matter what normal make one buys. The same cannot, however, be said of electronic calculators: not only is their layout variable but their *operational* keys do not necessarily correspond. For example, the valuable [EXC] key used in this chapter may not be present and one may have to use [CM][M+][C] instead, which only clears memory, transfers access to memory and then clears access, but does *not* put original contents of memory into access (or, alternatively, use [C][RM][CM] which does the reverse process, i.e. clears access, transfers memory to access and clears memory; one could, of course, write down access firstly on a piece of paper and reinsert this later).

2 (a) Several operations, such as [+] [−] [×] [÷], are put down *before* the relevant numbers, as will already have been noticed. For example:

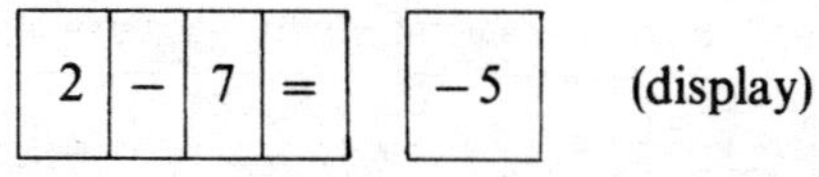

(display)

(b) Some do not operate until after the relevant number has been entered such as the keys [$1/x$][x^2][y^x]. For example:

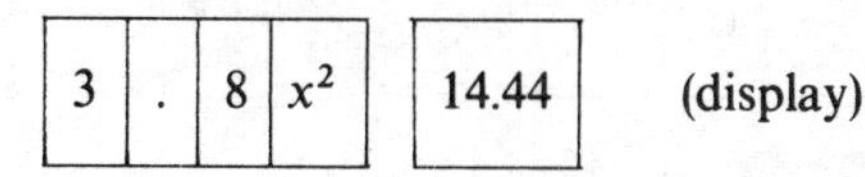

(display)

The first two of these do not need the use of 'equals' [=], but some of the keys carrying out these operations may need from 1 second to 2 seconds to complete them. It is unwise to press another key until each operation is finished.

3 Many calculators include keys for brackets, namely [(] and [)]. They have not hitherto been mentioned, as the present author has not often found the need for them in practical problems. Other keys seem to serve the purpose equally well and often yield shorter solutions. Look at this example:

Simplify $3\{4+2(7-1)\}$

Method 1: using brackets

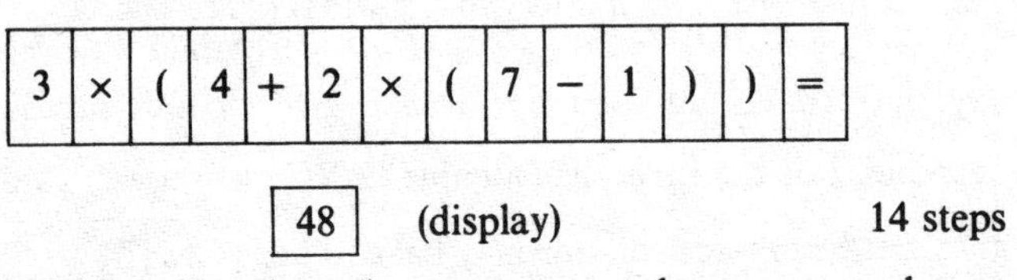

Method 2: without brackets (here we reverse the process and start from the inside)

7 − 1 = × 2 + 4 = × 3 = 48 12 steps

There are, however, cases in which brackets may be more suitable. Consider the case of two or more pairs of inner brackets, for example:

Simplify $5\{4.3(9.2-7.3)-6(2.9-0.08)\}$

Method 1: using brackets

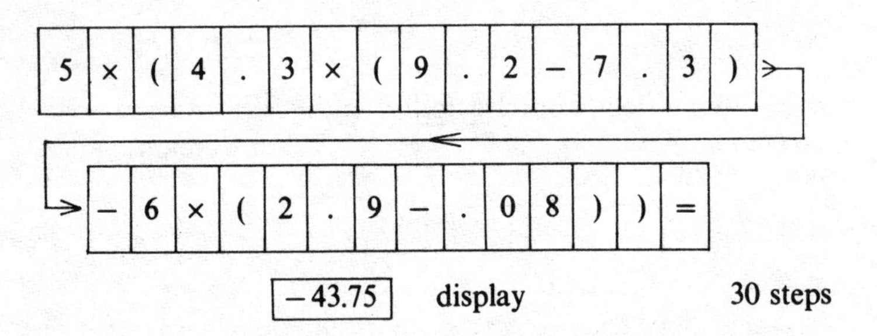

Method 2: without brackets

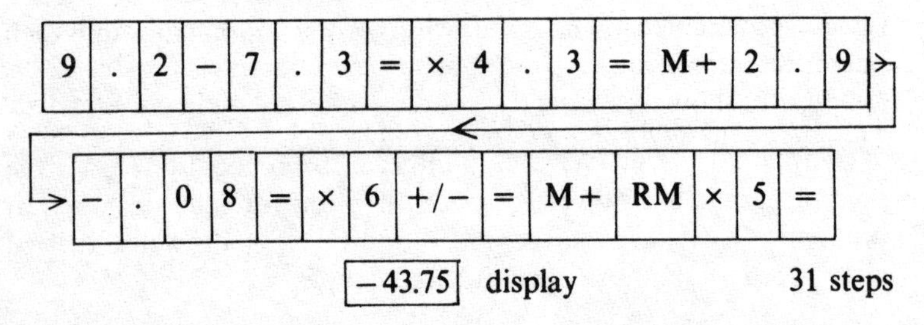

There is not much in it, but here, for the inexperienced, brackets are probably simpler to follow.

It will have been noticed that whereas we use different kinds of parentheses in writing, e.g. { }, ( ), [ ], the calculator only used one symbol ( ). This does not matter so long as we remember at which 'level' we are operating at any one time.

EXERCISE 3

Evaluate the following (a) using brackets, (b) without brackets (questions 1–5):

1 $14\{6(2-19)+81\}$ 2 $(6.3-0.08)(6.3+0.08)$

3 $\frac{19}{7}\{8.4+3.6(5.9-7.25)\}$ 4 $4.23\left\{\frac{1}{7}(5-2.8)-\frac{1}{8}(4+3.7)\right\}$

5 $\{3(7.05-9.21)-4(6.8-0.09)\}\{5(6.38+0.97)-8(4.72-7.09)\}$, to the nearest integer.

It may not be obvious that question 5 can be done without the use of brackets, nor is it a particularly elegant method, but one can proceed as follows:

(i) work out the left hand term as $3\times 7.05-3\times 9.21-4\times 6.8+4\times 0.09$; use Memory for final storage of -33.32,

(ii) work out the right hand term in the same way, but do not put this into the Memory, which is already occupied,

(iii) multiply (ii) by (i), thus | × | RM | = | .

6 Evaluate

$$\frac{(41.79-28.84)(17.8-61.95)}{(0.827-1.416)(7083-8119)}$$

using brackets and giving the result correct to three decimal places.

General note: There are frequently more ways than one of evaluating numerical expressions, whether simple or complicated. This book attempts to show some useful solutions which are rapid and effective. The reader, however, may well find some better ones!

9

Some Applications of the Electronic Calculator to Practical Problems

1 Choice of topics

In a book of this length, one can give a smattering of a wide variety of topics, with little skill being derived in any of them, or one can develop a more moderate number in greater depth. There are several other mathematics books in the Teach Yourself series. Rather than to duplicate what has been done in these companion volumes, it is better to select different material, *or*, where desirable, to give updated approaches thereto. The reader will find, in this chapter, some fairly detailed applications of electronic calculator technique with regard, firstly, to the mensuration of some simple curved surfaces and solids and, secondly, to some practical problems.

The aim here is to show the remarkable versatility of what is really quite a simple instrument, such as is illustrated in Fig. 63, when compared with some which are available. It is not an elaboration of keys which one needs, to solve many everyday problems, but sufficient skill to use a relatively small number efficiently. (One suspects that many of the more exotic instruments are bought by young people more as a prestige symbol than because they require all the functions which are present! Naturally this observation does not apply to those who are studying more advanced mathematical, scientific or engineering subjects.)

2 Mensuration of simple curved surfaces and solids

Before we look at curved surfaces and solids, we need to restate and extend the properties of the circle, which were given on page 101 (definitions 23–5).

Definition The word *circle*, insofar as mathematics is concerned, may be defined in two ways:

(a) A *circle* is a plane curve traced out by a point P moving so that it is at a constant distance from a fixed point O, which is called the centre of the circle (Fig. 64).

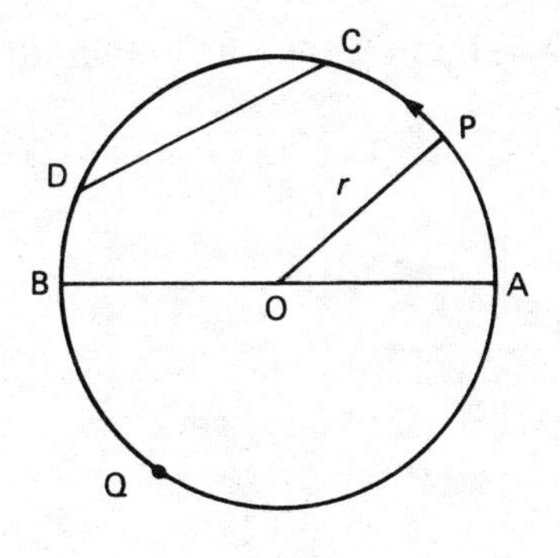

Fig. 64

The distance OP from O to P is called the *radius*, r.

(b) The whole area *within* the curve just defined is also called a *circle*, and when we use the word in *this* sense, we give the name *circumference* (of the circle) to the boundary curve itself. In Fig. 64, the curve PCDBQAP is the whole circumference; any part of it, such as AP or BQ, is called an *arc* of the circle.

More definitions A straight line drawn within a circle and terminating on the circumference at both ends, e.g. CD in Fig. 64, is a *chord*. If it passes through the centre O, e.g. AB, it is called a *diameter*, and as $OA = OB = r$ (the radius), it follows that a diameter is of length twice the radius.

Definition of π (Greek letter pi) The ratio *circumference* $\div$ *diameter* (of a circle) is constant. This constant ratio is called π. Its value is roughly $\frac{22}{7}$; more accurately 3.142, for use with four-figure logarithmic tables; for use with an electronic calculator, 3.1415927 is accurate enough for an eight-figure read-out. Many calculators now include a key for π, and this saves entering one of the above strings of numbers.

Hence we get $\qquad$ Diameter $(d) = 2r$,
and $\qquad$ Circumference $= \pi d = 2\pi r$,
and, of course, $\qquad$ Area $= \pi r^2$, which is familiar to many people.

Example 1 Find the circumference and the area of a circle of diameter 7.3 cm.
Take $\pi = 3.14159$ and give the results to three decimal places.

(i) Circumference πd

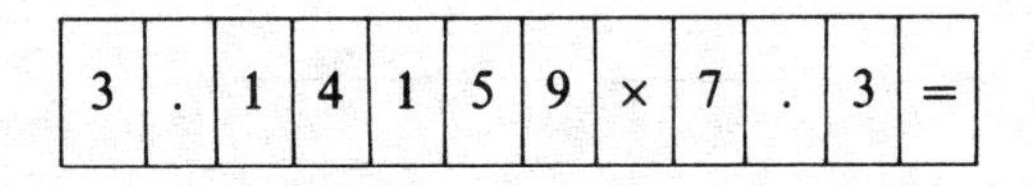

3	.	1	4	1	5	9	×	7	.	3	=

22.933 cm

(ii) Area πr^2

7	.	3	÷	2	=	x^2	×	3	.	1	4	1	5	9	=

41.853833 $\simeq$ **41.854 cm^2**

If a key for π is given on the calculator used, the above reduce to:

(i) Circumference

π	×	7	.	3	=

(ii) Area

7	.	3	÷	2	=	x^2	×	π	=

The same comment applies to later examples and questions. (*Now turn to Exercise 1, questions 1–3.*)

Definition A solid *right circular cylinder* is formed by revolving a rectangle through 360° about one of its edges (Fig. 65). In the diagram, rect. ABCD is revolved about AD, the axis of the cylinder; then AB ($= r$) is the radius of the cylinder and AD ($= h$) is its height. All cross-sections including end-sections are identical circles, each of area πr^2. The generating lines, such as BC, are all perpendicular to the ends. BC generates the curved surface.

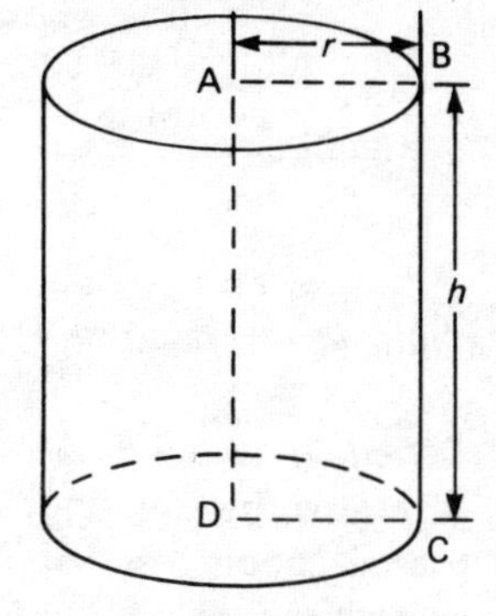

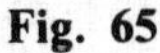
Fig. 65

The curved surface of the cylinder is of area $2\pi rh$ and the sum of the areas of the two end-sections is $2\pi r^2$.

$\therefore$ the total surface area is, on adding, $2\pi r^2 + 2\pi rh = 2\pi r(r+h)$.

The volume of the cylinder is $\pi r^2 h$. This corresponds exactly with the volume of a prism: see Chapter 7, Exercise 2 (definition and question 5) for

$$\pi r^2 h = (\pi r^2)h = \text{area of base} \times \text{height}.$$

Example 2 A solid right circular cylinder is of length 0.85 m. and diameter 6.4 cm. Calculate (i) its volume, in cm^3, (ii) its total surface area, in cm^2. Give both answers correct to one decimal place. If your calculator has no key for π, take $\pi = 3.1415927$.

We have $r = \frac{1}{2}d = 3.2$ cm, $h = 0.85$ m $= 85$ cm,

$\therefore$ volume $\pi r^2 h$ is given by:

$$\pi \times (3.2)^2 \times 85 \text{ cm}^3$$

i.e.

π	$\times$	3	.	2	x^2	$\times$	8	5	=

2734.4 cm³

and total surface area is:

$$2\pi rh(r+h) = 2\pi \times 3.2 \times (3.2 + 85) = (3.2 + 85) \times 2 \times \pi \times 3.2 \text{ cm}^2$$

3	.	2	+	8	5	=	×	2	×	π	×	3	.	2	=

1773.4 cm²

(*Now turn to Exercise 1, questions 1–6 and 18.*)

Definition A solid *right circular cone* is generated by rotating a right-angled triangle about one of its *legs* (i.e. a side which is *not* the hypotenuse). In Fig. 66, △AOB, which is right-angled at O, is rotated about the leg OA. Then AB generates the curved surface of the cone, and OB generates the circular base.

The slant height AB = l units; the base radius OB = r units and the *height* AO (as distinct from the *slant height*) = h units.

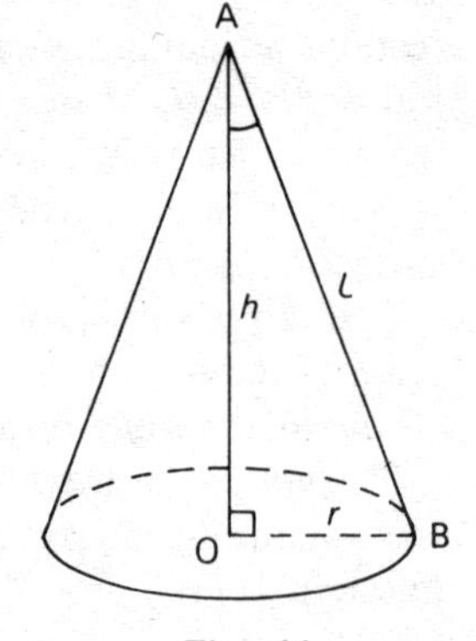

Fig. 66

We have

(i) the curved surface of the cone is of area πrl sq. units,
(ii) the base area is r^2 sq. units, and hence
(iii) the total surface area is $\pi r^2 + \pi rl = \pi r(r + l)$ sq. units.

By Pythagoras' theorem, applied to AOB, as angle AOB = 90°:

$$l^2 = r^2 + h^2,$$

hence the area of the curved surface can be expressed in terms of base radius and height (instead of slant height); it is $\pi r\sqrt{r^2 + h^2}$, only the positive root being of interest here. The corresponding total surface area is

$$\pi r(r + \sqrt{r^2 + h^2}).$$

Finally, the volume is $\frac{1}{3}\pi r^2 h$. This corresponds exactly with the volume of the pyramid i.e. it is of the form

$$\frac{1}{3}(\text{Area of Base}) \times (\text{Height}), \text{ for } \frac{1}{3}\pi r^2 h = \frac{1}{3}(\pi r^2)h.$$

Example 3 The slant height of a right circular cone is 29 cm, and its base radius is 20 cm. Find its curved surface and its volume, correct to the nearest integer.

We have curved surface is $\pi rl = \pi \times 20 \times 29\ \text{cm}^2$

i.e.

π	$\times$	2	0	$\times$	2	9	=

1822 cm² (nearest sq. cm.)

As $r^2 + h^2 = l^2$, then $h^2 = l^2 - r^2$; $\therefore\ h = \sqrt{l^2 - r^2}$

so the volume $\frac{1}{3}\pi r^2 h = \frac{1}{3}\pi r^2 \sqrt{l^2 - r^2} = \sqrt{29^2 - 20^2} \times \pi \times 20^2 \div 3$

2	9	x^2	−	2	0	x^2	=	$\sqrt{x}$	$\times$	π	$\times$	2	0	x^2	$\div$	3	=

8796 cm³

Note: It is best to work out $\sqrt{29^2 - 20^2}$ first. Had we used the Memory, however, we could have evaluated the result in the original order, but more keys would have been required. (Incidentally, $\sqrt{29^2 - 20^2} = 21$.) (*Now turn to Exercise 1, questions 7–10.*)

Definition As with the cylinder and cone, a *sphere* has two definitions, dependent upon whether one considers it to be entirely hollow (i.e. as being only a surface) or as completely solid (Fig. 67).

(a) A *sphere* is the locus of a point P in space, moving so that it is at a definite distance from a fixed point O. The distance OP ($= r$) is the radius of the sphere.

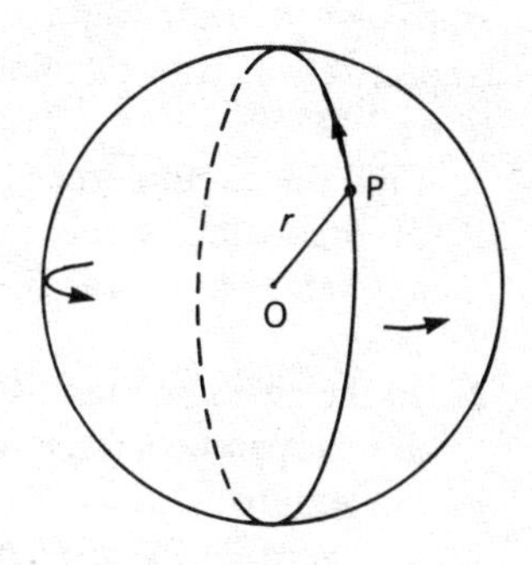

Fig. 67

(b) A *sphere* is a solid body *bounded by* a surface which is made up of all points which are at a definite distance r from a fixed point O. It follows that (b) is bounded by (a).

The surface area of a sphere is $4\pi r^2$ sq.units.

The volume (already used in Chapter 8, Example 6) is $\frac{4}{3}\pi r^3$ cu.units.

Example 4 *Assuming that the Earth is a sphere* of mean diameter 7913 miles, calculate its total surface area. (The Earth is really an oblate spheroid, being slightly flattened towards the poles: equatorial diameter $\simeq$ 7927 miles; polar diameter $\simeq$ 7900 miles.) Give the result in standard form, correct to two decimal places.

$$\text{Surface area} = 4\pi r^2 = \left(\frac{7913}{2}\right)^2 \times \pi \times 4$$

7	9	1	3	÷	2	=	x^2	×	π	×	4	=

1.9671 08

$= 1.9671 \times 10^8 \simeq$ **1.97×10^8 sq. miles**

(which, in full, is 197 000 000 square miles).

Incidentally, the result of our assumption that the Earth is a sphere is not bad. Our estimate was about 196.7 million square miles; a more accurate figure than our mean value approximation has been found to be 196.9 million. Both give the result 1.97×10^8, correct to two decimal places. (*Now turn to Exercise 1, questions 11–17*)

EXERCISE 1

In all the following questions, where relevant, take the words *cylinder* and *cone* to mean right circular cylinder and right circular cone, respectively. If your calculator does not have a key for π, take its value to be 3.14159.

1 Find the circumference of a circle of diameter 9 cm, giving the answer correct to two decimal places.

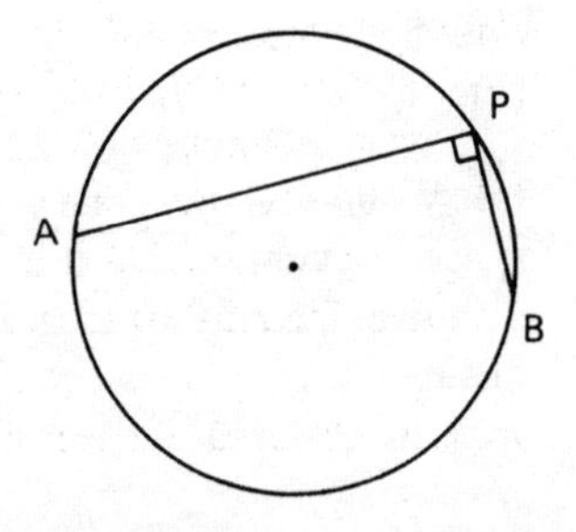

2 In the given diagram, PA and PB are perpendicular chords of the circle shown. PA = 18 cm and PB = 7.5 cm. By referring to Theorem 9 (converse) in Chapter 6, and using your calculator, find

the length of a diameter. Hence find the area of the circle, to two decimal places.

3 Find the circumference of a circle which has an area of $100\,\text{cm}^2$, giving the result correct to two decimal places.

(*Hint*: $\pi r^2 = 100 \Rightarrow r^2 = \dfrac{100}{\pi}$; then find the value of r from $\boxed{\sqrt{x}}$ key. After that, it is plain sailing.)

4 Calculate the area of the curved surface of a cylinder of height 12 cm and diameter 3 cm, giving the answer correct to one decimal place.

5 A drinking chocolate tin is of diameter 10 cm and height 10.9 cm. Find its volume to the nearest cubic centimetre. (This is an actual tin, which is designed to hold 500 grams of chocolate.)

6 A rectangular sheet of metal of length 30 cm and height 23 cm is to be used to form the curved surface of a cylindrical can of the same height. After it has been rolled into shape, the 23 cm edges are soldered together. A circular base is to be cut out of another piece of metal and soldered to one end of the hollow cylinder. What is its radius? What is the volume of the contents which can be kept in the tin when it is stood on its base?

7 Find the volume of a cone of height 7.8 cm and base radius 2.6 cm, correct to one place of decimals.

Calculate also the area of the curved surface, to the same accuracy.

8 A solid cone is of slant height 13 cm and base radius 5 cm. Find (i) its height, (ii) its volume, (iii) its total surface area.

9 A solid cone is to have a curved surface area of $1000\,\text{cm}^2$. The radius of its base is to be 9.84 cm. Find (i) the slant height, (ii) the height, (iii) the volume, of the cone, correct to the nearest whole number.

10 The diagram depicts a metal tank made in the form of a hollow cylinder welded on to the top of an inverted cone of the same basal diameter. There is an end plate at BC but not, of course, at AD. The total height of the tank, which has the end BC horizontal and vertex V downwards, is 12 ft. The cylindrical part is of height 7 ft and diameter 6 ft. Calculate

(i) the area of sheet metal used to make the tank, to the nearest ft^2,

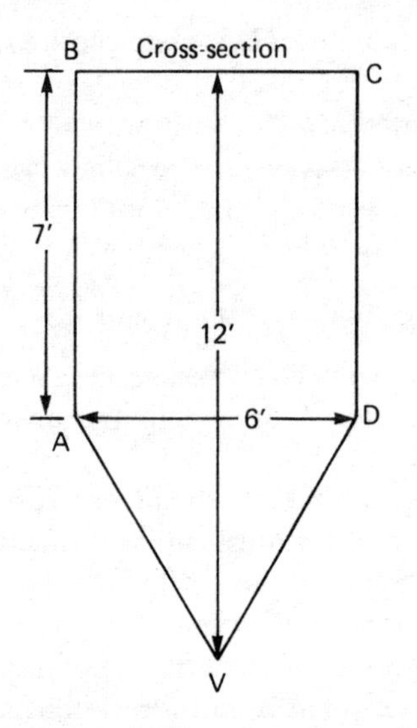

(ii) the volume of water which the tank will hold, to the nearest ft^3,

(iii) assuming that there is a tap, at V, through which water can flow at a rate of 1 gallon every 10 second, how long it will take to empty a tankful (take 1 cu.ft. = 6.23 gallons).

11 A sphere has a volume of $20\,cm^3$. What is the volume of a sphere of one-third of the radius? (Is a calculator needed?)

12 Find the surface area of a sphere of diameter 17.3 cm (correct to one decimal place).

13 Determine the volume of a sphere of radius 4.823 cm (correct to one decimal place).

14 Find the *total* surface area of a solid hemisphere (half-sphere, cut off by a plane through the centre of the sphere) of radius 8.6 cm.

Calculate the volume of the hemisphere.

If its density (mass per unit volume) is 7.86 grammes per cm^3, find the total mass of the hemisphere, giving the answer in kilogrammes, correct to the nearest 0.1 (1 kilo = 1000 gm).

15 A boiler is made in the form of a cylinder, of diameter 2.15 m, having hemispherical ends. The cylindrical part is of length 4.28 m.

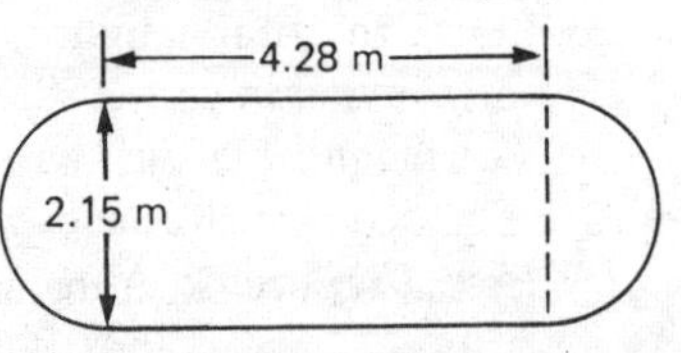

Calculate, correct to one decimal place in each case, neglecting the thickness of the metal,

(i) the volume of the boiler, in m^3,

(ii) the total external surface area, in m^2.

16 A steel ingot in the form of a cuboid 0.75 m by 0.2 m by 0.1 m is melted down to form ball-bearings of diameter 0.4 cm. How many ball-bearings of the required specification can be made? (To the nearest 100.)

17 Find the volume of a solid hemisphere, of which the total surface area is $64\,cm^2$ (two decimal places).

18 A large cylindrical hollow pipe is to be laid in the ground. Its length is 5 m and its external and internal radii are 12.4 cm and 11.4 cm respectively. Find the volume of material used in making the pipe, giving the result in cubic metres, correct to 4 decimal places ($1\,m^3 = 10^6\,cm^3$).

The pipe is made of metal of mass 7400 kg per cubic metre. Find the mass of the pipe, in tonnes, correct to the nearest 0.1 tonne (1 tonne = 1000 kg).

19 In Chapter 7, page 126, the volume of a regular pyramid on a square base is given by:

$$V = \frac{4}{3}a^2\sqrt{b^2 - 2a^2*}, \quad *(\text{N.B. } \textit{Not } (2a)^2)$$

where $2a$ is the length of a base edge and b is the length of a slant edge.

(i) Calculate the volume of the pyramid shown, in m^3.

(ii) Find the *formula* for the area of the slant face VAB in terms of a and b.

(iii) Calculate the total slant area of the pyramid, in m^2.

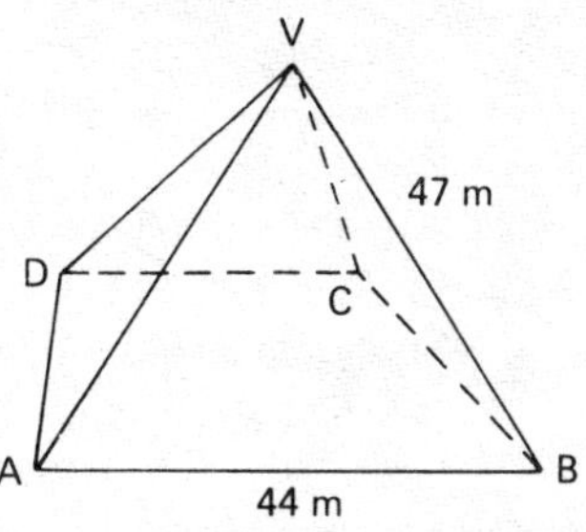

(Give the results for (i) and (iii) correct to the nearest integer)

20 A cone is to have a height of 7.8 cm and a volume of $250\,cm^3$. Find the diameter of its base, correct to two decimal places.

21 The diagram shows a circular paper disc of radius r. Part of the disc (the sector AOB, which is shaded, and in which angle

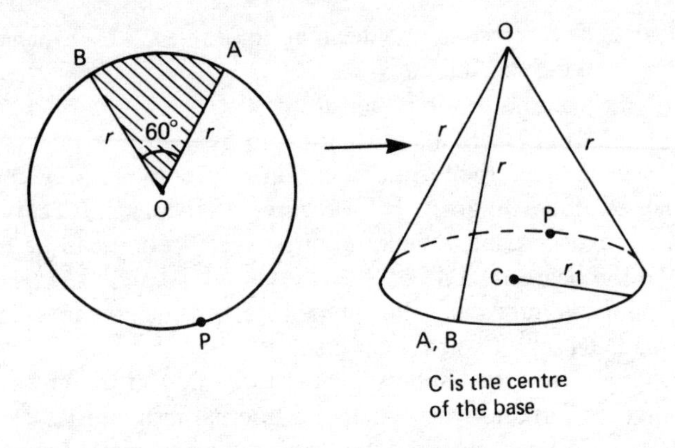

AOB = 60°) is cut away. The remainder is reshaped as a hollow cone with vertex O, the lines OA and OB now coinciding. In terms of r, (i) what is the radius r_1 of the base, in terms of r; (ii) what area of paper is required to make the (circular) base for the cone?

If $r = 8$ cm, find the area of the curved surface of the cone, correct to one decimal place.

10

Calculations Concerned with Financial Problems

1 Introduction

We shall now be taking a look at a number of monetary problems, the resolution of which may be greatly simplified by the use of a good electronic calculator.

The background, theory and formulae relating to profit and loss, simple and compound interest, and investments of certain types are covered in *Arithmetic* (Teach Yourself Books), Chapters 9, 14 and 16, respectively. These topics are included below, rather more briefly described than was the case in *Arithmetic*, but with all necessary formulae explained in suitable form for solution by calculator.

The more difficult section on Mortgages and Loans, which follows those just mentioned, does not appear in the above-named book, but I have evolved a practical formula which is given in Section 6.

2 Percentage key: [%]

The following examples illustrate the basic use of this key. In some calculators, to complete the problem the [=] sign must be depressed after [%]. We shall take this case, but there are machines wherein this last step is unnecessary. Incidentally, per cent (Latin, *per centum*) means 'by the hundred', e.g. $\frac{3}{5} = \frac{60}{100} = 60\%$.

Example 1 Express the following as percentages (to two decimal places, where necessary): (i) $\frac{3}{8}$, (ii) $\frac{17}{23}$, (iii) $1\frac{4}{7}$, (iv) 0.89.

(i)

3	÷	8	%	=

37.5%

(ii)

1	7	÷	2	3	%	=

73.91% (two decimal places)

(iii) $1\frac{4}{7} = \frac{11}{7}$;

1	1	÷	7	%	=

157.14% (two decimal places)

(iv) A calculator is not needed: $0.89 = 0.89 \times 100\% =$ **89%**, but using the method of (i)–(iii) above gives

.	8	9	÷	1	%	=

89%

Example 2 What is the value of (i) $14\frac{1}{2}\%$ of £172, (ii) 23% of $4\frac{1}{4}$kg?

(i)

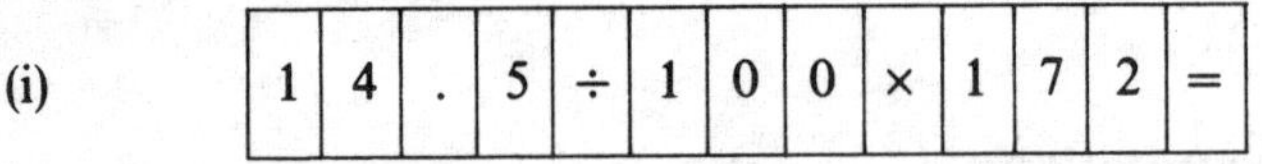

£24.94

(ii) It is *shorter* work if we mentally (when convenient) move the decimal point two places to the left before using the calculator: 23% $= \frac{23}{100} = 0.23$.

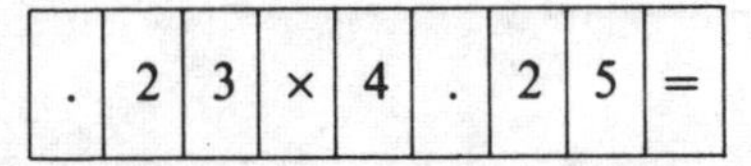

$0.9775 \simeq$ **0.98 kg** (two decimal places)

EXERCISE 1

1 Express the following as decimals (a calculator is not really needed in this question, but may be used for practice):
(i) 25% (ii) 36.4% (iii) 7.5% (iv) 253%

2 Express the following numbers as percentages of 1 (to 4 dec. pl. where needed):

(i) $\frac{1}{17}$ (ii) $\frac{31}{248}$ (iii) 0.836 (iv) 5.02 (v) $\sqrt{2} \div \sqrt{3}$

3 By how much per cent is $\frac{27}{43}$ greater than $\frac{38}{61}$, correct to two decimal places? (Be careful!)

4 Find the value of
(i) $38\frac{1}{2}$% of £942 (ii) 128% of 4.85 litres, in centilitres
(iii) $3\frac{1}{4}$% of £8.

5 Express £247.19 as a percentage of £3955.04.

3 Profit and loss

For problems of this type the calculations are very simple, provided that one first sets out, mentally or on paper, Cost Price (C.P.), Selling Price (S.P.) and Profit, correctly related.

Let the *percentage* profit be p, then

$$p = 100 \times \frac{\text{Profit}}{\text{C.P.}} \quad \text{i.e.} \quad \text{Profit} = \frac{p}{100} \times \text{C.P.} \quad \ldots\ldots (1)$$

Now

$$\text{Profit} = \text{S.P.} - \text{C.P.} \quad \ldots\ldots (2)$$

From (1) and (2)

$$\text{S.P.} - \text{C.P.} = \frac{p}{100} \times \text{C.P.}$$

i.e.

$$\text{S.P.} = \left(1 + \frac{p}{100}\right) \times \text{C.P.}$$

$\therefore$

$$\text{S.P.} = \frac{100 + p}{100} \times \text{C.P.} \quad \ldots\ldots (3)$$

and

$$\text{C.P.} = \frac{100}{100 + p} \times \text{S.P.} \quad \ldots\ldots (4)$$

Example 3 A vase is bought for £2.75 and sold for £3.32. What is the percentage profit? (Two decimal places.)

$$p = 100 \times \frac{\text{Profit}}{\text{C.P.}} = £100 \times \frac{3.32 - 2.75}{2.75}\%$$

The neatest way is

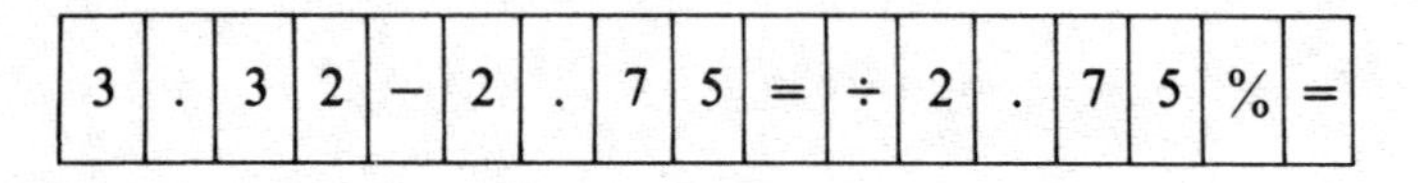

20.73% (two decimal places)

Example 4 Brown purchases one dozen spades for a *total* cost of £47. He aims to sell them at 35% profit. How much should he charge, correct to the nearest 1p, for each spade, (a) if there is no Value Added Tax (VAT), (b) if VAT at 15% is included in the selling price?

The cost of one spade is $£\frac{47}{12}$.

$$\text{C.P. } 100, \text{ S.P. } 135 \Rightarrow \text{S.P.} = \frac{135}{100} \times \text{C.P.} = £\frac{135}{100} \times \frac{47}{12}$$

∴ the cost price of one spade, excluding VAT, is given by

1	3	5	÷	1	0	0	×	4	7	÷	1	2	=

5.2875 ≏ **£5.29**

(The working is shortened if one puts 1.35 instead of 135 ÷ 100)
The selling price of one spade, including VAT and starting from scratch,

is $$\text{S.P}_2. = £\frac{135}{100} \times \frac{47}{12} \times \frac{115}{100} = £1.35 \times \frac{47}{12} \times 1.15$$

i.e.

1	.	3	5	×	4	7	÷	1	2	×	1	.	1	5	=

6.080625 ≏ **£6.08**

Example 5 An article is sold for £17.30 at $27\frac{1}{2}\%$ profit. What was the cost price?

$$\text{C.P.} = \frac{100}{100+p} \times \text{S.P.} = £\frac{100}{127.5} \times 17.30$$

A neat way is

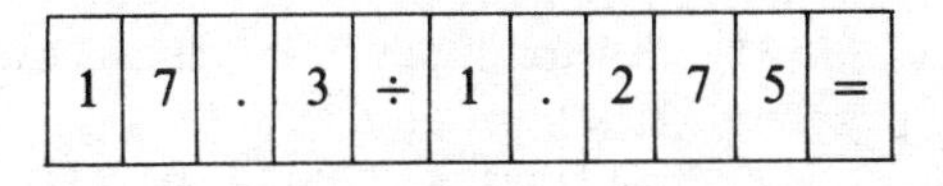

£13.57 (to nearest 1p)

Example 6 A tea service is bought in two instalments. The first is £108 and represents 45% of the total cost. How much is the second instalment?

First instalment 45% ∴ second instalment is $(100-45)\% = 55\%$

∴ Second instalment is $£\frac{55}{45} \times 108$

5	5	÷	4	5	×	1	0	8	=

£132

(The total cost was £108 + £132 = £240.)

EXERCISE 2

1 What is the selling price, if the cost price is £21 and the profit is 34%?

2 Find the selling price, if an article costing £10.75 is sold at a loss of 8%.

3 Find the percentage profit or loss when:
(i) the cost price is £83.50 and the selling price is £112.25,
(ii) the cost price is £23.46 and the selling price is £22.87.

4 If an article is sold for £119.68 at a profit of 28%, what was the original cost price?

5 A tin of red salmon costs £2.65 and sells for £3.71. What is the percentage profit?

6 A clock which cost £76 is intended to be retailed at 45% profit.

During a sale the retail price is reduced by 12%. What is the sale price, correct to the nearest 5p?

7 Williams purchases 60 jars of coffee for a total of £44.40, and he wishes to make $28\frac{1}{2}$% profit. What will he charge for one jar?

8 Roberts buys six identical electric kettles for a *total* price of £57. He wishes to sell them for a *net* profit of £5 on each one. If he is obliged to add 15% VAT to his net selling price, how much will a customer pay for a kettle?

9 Brown buys a secondhand car for £1280 and prices it to sell at 35% profit. The end of the year is, however, approaching and he suddenly decides to reduce his whole stock of secondhand cars by 10% in price. (i) If he now sells this car, what *actual* profit does he make on this particular transaction? (ii) What *percentage* profit does he make? (iii) Does the cost price affect the percentage profit?

4 Simple interest

If £P is the principal (i.e. the sum of money loaned or invested), r% p.a. is the rate of interest *per cent per annum* (i.e. the number of pounds sterling paid as interest on each £100 loaned or invested *for each year*), and n is the number of years for which the principal remains constant, then the simple interest (I_s) is given by the formula:

$$I_s = \frac{\text{Principal} \times \text{rate per cent} \times \text{number of years}}{100} = \frac{Prn}{100} \quad \ldots (1)$$

where Prn means, of course, $P \times r \times n$

Example 7 Find the simple interest on £2430 invested at 8.5% for $2\frac{3}{4}$ years.

Assuming this is to be free of tax, we have $P = 2430$, $r = 8.5$, $n = 2.75$.

$$\therefore \quad I_s = \frac{2430 \times 8.5 \times 2.75}{100} = \text{(mentally) } 24.3 \times 8.5 \times 2.75,$$

saving four keys,

2	4	.	3	×	8	.	5	×	2	.	7	5	=	568.0125

i.e. the simple interest is **£568.01** (to the nearest 1p).

Inverse simple interest problems

The simple interest formula (1) above contains four unknown quantities I_s, P, r, n. We have obtained I_s in terms of P, r, and n, but we can rearrange the formula to give any one of the four unknowns in terms of the other three.

We have

$$I_s = \frac{Prn}{100} \quad \ldots\ldots\ldots\ldots (1)$$

$$\therefore \quad 100\, I_s = Prn$$

i.e. on changing sides $Prn = 100\, I_s$ . . . . (1a)

Divide both sides of (1a) by rn

$$\therefore \quad P = \frac{100\, I_s}{rn} \quad \ldots\ldots\ldots\ldots (2)$$

Divide both sides of (1a) by Pn

$$\therefore \quad r = \frac{100\, I_s}{Pn} \quad \ldots\ldots\ldots\ldots (3)$$

Divide both sides of (1a) by Pr

$$n = \frac{100\, I_s}{Pr} \quad \ldots\ldots\ldots\ldots (4)$$

Example 8 What principal is needed to bring in a simple interest of £480 in $1\frac{1}{4}$ years at 7.75%, free of tax?

We have $I_s = £480$, $r = 7.75$ and $n = 1.25$. Using (2) above

$$P = \frac{100\, I_s}{rn} = £\frac{100 \times 480}{7.75 \times 1.25}$$

1	0	0	×	4	8	0	÷	7	.	7	5	÷	1	.	2	5	=

4954.8387 =**£4954.84**

Example 9 How long will it be before a sum of money will be trebled at 9.25% p.a. simple interest, free of tax? (Give the answer in years and days.)

We have no statement as to the principal involved, which we therefore take as P, in which case $I_s = 3P - P = 2P$. (After deducting the principal from the final total amount.) Also $r = 9.25$.

$$\therefore \text{ from (4) above, } n = \frac{100\,I_s}{Pr} = \frac{100 \times 2 \times \overset{1}{\cancel{P}}}{9.25 \times \underset{1}{\cancel{P}}} = \frac{200}{9.25}$$

2	0	0	÷	9	.	2	5	=

21.621622 (write down 21 years)

−	2	1	=	×	3	6	5	=

226.89189 (write down 227 days)

The time required is **21 years 227 days.**

This presupposes the final year is not a leap year. If it were, what would be the difference, in days?

EXERCISE 3

1 Find the simple interest on
  (i) £3640 at 10.5% for three years,
  (ii) £2593.56 at $8\frac{3}{4}$% for 5 years 4 months.
2 Peter invests £680 at 9% simple interest. He withdraws the interest when it reaches £306. For how long was it invested?
3 What sum of money, to the nearest £1, must be invested in order to bring in a simple interest of £800 at $7\frac{1}{2}$% for $3\frac{1}{2}$ years?
4 Smith loans a sum of money on which he requires back exactly twice the loan, including simple interest, at the end of eight years. What percentage interest is he charging?
5 Thomas puts £760 into a building society which pays him 11.25% interest at the end of the year. He adds this interest to the capital. Assuming that the rate of interest remains unchanged, how much interest will he receive for the following year? (This is the basic idea of compound interest – see the next section.)

5 Compound interest

Let us consider how the capital (or principal) builds up from a single sum of money, £P, invested in a scheme which pays interest at r% once yearly if the interest is automatically added to the capital, i.e. it is *compounded*. (This is *not* the case with simple interest.)

Interest each year is r%, i.e. $\frac{r}{100}$ of principal

Thus to each £1 a sum of £$\frac{r}{100}$ is added at the end of each year.

Hence, at the end of one year, the *amount* on £1 is £$\left(1+\frac{r}{100}\right)$.

At the end of two years, it has become

$$£\left(1+\frac{r}{100}\right)+£\left(1+\frac{r}{100}\right)\frac{r}{100}=£\left(1+\frac{r}{100}\right)\left(1+\frac{r}{100}\right)$$

$$=£\left(1+\frac{r}{100}\right)^2.$$

Proceeding thus, at the end of n years, the amount on £1 is £$\left(1+\frac{r}{100}\right)^n$.

$\therefore$ on £P the amount is $\quad £P\times\left(1+\frac{r}{100}\right)^n$

If the amount on £P for n years is called A_n, we have

$$A_n=£P\left(1+\frac{r}{100}\right)^n$$

The compound interest, I_c, is given by

$$I_c=A_n-P$$

i.e. the compound interest is the amount accumulated *less* the capital, or principal, initially paid in.

Example 10 Jones invests £1840 in a building society, the money being left to accumulate for $3\frac{1}{2}$ years. Interest is at the rate of 10% *per annum*, payable *half-yearly*. Find the amount at the end of the period and the total interest received.

We have $P = 1840$, $n = 7$ (half-years), $r = \frac{1}{2} \times 10\% = 5\%$ (half-yearly).

$$A = 1840\left(1 + \frac{5}{100}\right)^7 = 1840\,(1.05)^7$$

At this stage we use the method shown on page 148 (Example 8).

Amount A_n

1	.	0	5	y^x	7	×	1	8	4	0	=

2589.0648 **£2589.06**

Interest I_c →

−	1	8	4	0	=

749.06477 **£749.06**

It is interesting to compare this compound interest with simple interest on the same principal, at the same rate and for the same period of time.

We know that the simple interest is $\frac{Prn}{100}$.

Now $P = £\,1840$, $r = 10$, $n = 3.5$ (there is no need to subdivide r and n for simple interest), so the simple interest is

1	8	4	0	×	1	0	×	3	.	5	÷	1	0	0	=

£644

∴ Compound Interest (I_c) exceeds Simple Interest (I_s) by

$I_c - I_s$ (i.e., on continuing)

−	7	4	9	.	0	6	=	+/−

£105.06

Example 11 Black buys £2845 of $13\frac{1}{4}\%$ Treasury Loan (1997), interest being payable half-yearly in January and July. If tax has already been deducted at 30 % of the interest paid on each occasion, how much does Black receive each half-year?

The stock cost him £2872.45, including expenses. What is the *net* (after tax) percentage yield on outlay?

Gross *annual* interest is $£2845 \times \frac{13\frac{1}{4}}{100}$

Tax is deducted at source at 30 %, so Black receives 70 % $= \frac{70}{100}$ of gross interest.

∴ Half-yearly income is

$$£\frac{1}{2} \times 2845 \times \frac{13.25}{100} \times \frac{70}{100}$$

$$= £\frac{2845 \times 13.25 \times 70}{2 \times 10^4}$$

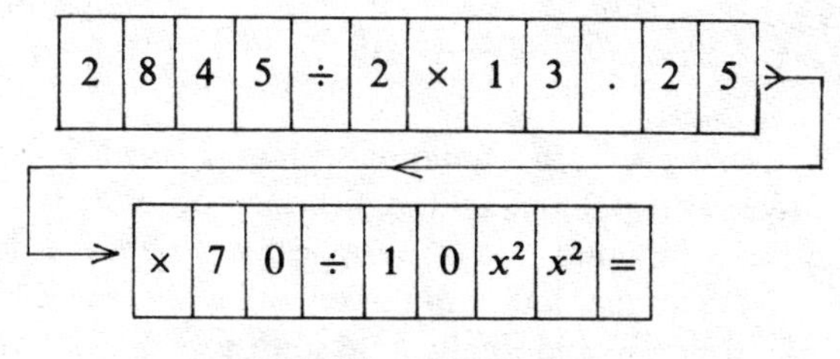

131.93688 ≏ **£131.94**

The net percentage yield on outlay is the net interest for the *whole* year, divided by the total cost of the stock, and then expressed as a percentage, i.e.

$£\frac{131.93688 \times 2}{2872.45} \times 100\%$ ⟵ usually done by [%] [=]

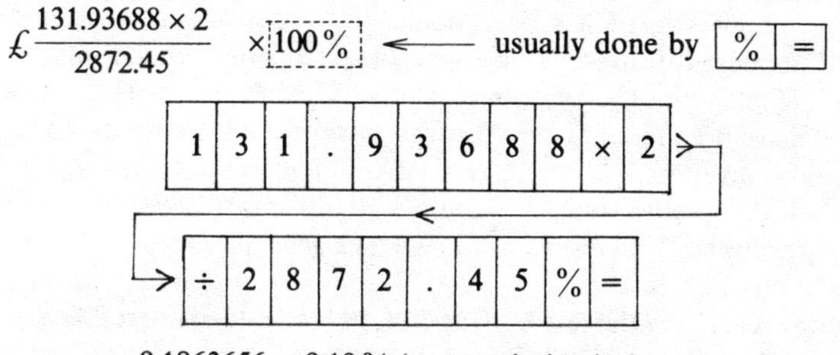

9.1863656 ≏ 9.19 % (to two decimal places)

(*Note*: The multiplication of 131.93688 by 2 needs to be done *before* dividing by 2872.45 so as to ensure that we have *one* number in the numerator and *one* in the denominator prior to pressing the percentage key. By doing this we effectively get 263.87376 ÷ 2872.45, as needed.)

EXERCISE 4

Find the *amount* and the *compound interest* on each of the following investments (questions 1 – 4): use $\boxed{y^x}$ key where desirable (see page 148, Example 8, and pages 173–4, Example 10).

	Principal	*Rate % per annum*	*No. of Years*	*Interest Yearly or Half-Yearly*
1	£ 750	6	2	Yearly
2	£ 3460	$7\frac{3}{4}$	3	Yearly
3	£ 4903.68	$9\frac{1}{2}$	4	Half-Yearly
4	£17843.73	$11\frac{1}{4}$	$3\frac{1}{2}$	Half-Yearly

5 Clark invests £4655 in a building society paying 8.5 % per annum in half-yearly instalments. Find, for a period of $4\frac{1}{2}$ years, by how much the compound interest (applicable if he leaves his interest payments to accumulate with the capital) would exceed the simple interest (which would apply if he were to withdraw the interest payments at the times when they were paid).

6 A building society is paying 10.5 % per annum, interest being paid half-yearly, tax having been deducted at the standard rate. Green invests £3850 with this society, intending to leave the money to accumulate, with interest, for exactly five years. Just two years after the investment has been made, the society reduces its annual interest rate to $8\frac{1}{2}$%, because of a reduction in the minimum lending rate. Calculate by how much Green's expected interest will have been reduced over the five-year period, if he decides to leave his investment standing. (*Hints*: Although this may, at first, look a little disconcerting, it is quite straightforward if we look at the compound interest formula. We take general cases for the rates of interest:

Let the initial rate be r_1 and the revised rate of interest be r_2, for each *full* year. Green would originally have expected the amount for 5 years to have been $£P\left(1+\frac{r_1}{200}\right)^{10}$; he must now be content with $£P\left(1+\frac{r_1}{200}\right)^4\left(1+\frac{r_2}{200}\right)^6$. The difference between these is the reduction in interest, after putting $r_1 = 10.5$, $r_2 = 8.5$, because Amount = Principal + Interest (see earlier), and the principal is unchanged,

$$\therefore A_1 - A_2 = P + I_1 - (P + I_2) = P + I_1 - P - I_2 = I_1 - I_2$$

where $A_1 = 3850\,(1.0525)^{10}$ and $A_2 = 3850\,(1.0525)^4\,(1.0425)^6$. In some calculators, it may be necessary to calculate A_2 firstly.

7 Evans buys £2000 of $12\frac{1}{2}\%$ Exchequer Stock (1994) at $94\frac{1}{8}\%$. What gross annual interest will he receive?

If the total cost of the purchase *including* brokerage, VAT on the brokerage, and contract stamp, was £1897.48, what is the *net* annual yield % on outlay, after deducting 30% tax? (Interest is paid half-yearly and tax will have been deducted at source, but it is sensible to check income from investments.)

6 Mortgages and loans involving regular repayments: $\boxed{y^x}$ key desirable

Although the mathematics behind this topic is rather more advanced, it may be of considerable interest to many readers, for the following information gives some idea of one way in which estimations may be made in connection with mortgages. The letters used need careful definition and the calculations must be adjusted each time there is a change in the rate of interest and/or a variation in the payments made. Although for simplicity, the mortgage repayments are initially taken as though made once yearly and hence the capital sum on loan (i.e. the principal) is adjusted only annually, the *same formula* will do whether for a six-monthly, three-monthly, one-monthly or even shorter period, but the value of n (the number of repayments) increases accordingly and the value of r (the rate per cent for the time-interval between repayments) changes in inverse proportion to n, i.e. r varies as $1 \div n$.

(i) Let P_0 be the initial loan (the principal on day 1 of year 1), then $P_1, P_2 \ldots\ldots P_n$ are the sizes of the outstanding loan at the ends of years 1, 2 .... n respectively.

(ii) £E is the value by which the annual repayment to the building society *Exceeds* the initial interest due for the first year. (The formula given below presupposes that the annual repayments remain unchanged.) The importance of E lies in the fact that the first year is critical. If $E = 0$, for example, the mortgage is never paid off, for the mortgagee is only just paying the interest due and he is not repaying any of the money he has borrowed; but if $E > 0$, the borrower, by the end of the first year, will have paid off a little of the loan. In consequence, at the end of the second year he will have rather less interest to pay. Thus repayment of the loan becomes increasingly rapid as time goes by. Fig. 68 shows the changing situation for a loan of £16 000 taken out for a period of twenty years. (The diagram is not to scale.)

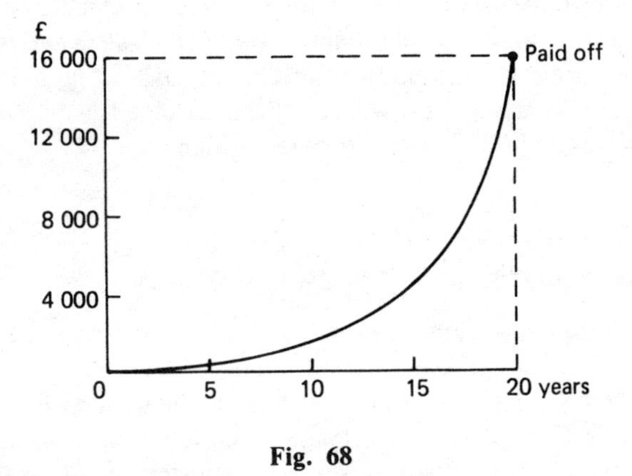

Fig. 68

(iii) r is the rate % per annum, hence the interest on £P for 1 year at r%, i.e. at $\frac{r}{100}$, is £$\frac{Pr}{100}$. It follows that $E > \frac{Pr}{100}$, from the above.

(iv) n is the number of complete years.

The following formula gives £P_n the size of the loan still outstanding after n years have been completed, in terms of P_0, E, r and n:

$$P_n = P_0 - \frac{100E}{r}\left\{\left(1+\frac{r}{100}\right)^n - 1\right\}$$

KEYS PRESSED IN ORDER	NUMBERS ON DISPLAY	EXPLANATION EACH PROCESS IS DELAYED UNTIL A SUITABLE KEY SUBSEQUENT TO THE OPERATION CONCERNED IS DEPRESSED
1 . 1 4 y^x	1.14	
1 0	10	
−	3.7072213	$(1.14)^{10}$ completed *or* (without y^x) 1 . 1 4 log × 1 0 = INV log (see page 182)
1	1	
= ×	2.7072213	$(1.14)^{10} - 1$ completed
2 0 0 0 0	20000	
÷	54144.426	$20\,000\{(1.14)^{10} - 1\}$ completed
1 4 +/−	−14	Number required is negative
+	−3867.459	$-\frac{20\,000}{14}\{(1.14)^{10} - 1\}$ completed
1 0 0 0 0	10000	
=	6132.541	$10\,000 - \frac{20\,000}{14}\{(1.14)^{10} - 1\}$ result.

Although this looks perhaps a little unwieldy, it is remarkably amenable when an electronic calculator is employed. Let us take the simple case of a loan of £10 000 at 14% p.a., the *annual* payments paid by the borrower being constant at £1600. How big is the outstanding loan at the end of ten years?

First-year interest due is:

$$\frac{P_0 r}{100} = \frac{10\,000 \times 14}{100} = 1400$$

$$\therefore \quad E = 1600 - 1400 = 200$$

$$P_{10} = 10\,000 - \frac{100 \times 200}{14}\left\{\left(1 + \frac{14}{100}\right)^{10} - 1\right\}$$

$$= 10\,000 - \frac{20\,000}{14}\left\{(1.14)^{10} - 1\right\}$$

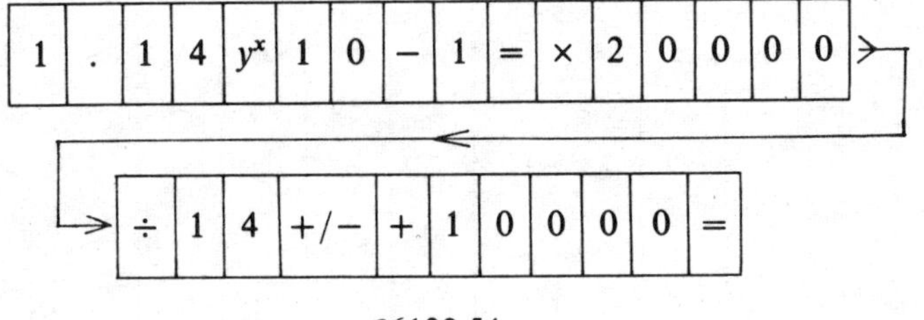

£6132.54,

and this is the size of the loan outstanding at the end of 10 years.

An analysis of the display on page 179 may be helpful to the reader in that it shows how the calculation proceeds.

For someone inexperienced in calculations of this type, it is disconcerting to find an earlier part of the calculation appearing after having started the next part, but one soon gets used to the idea.

Example 12 Edwards takes out a mortgage of £14 000 at an interest rate of 11% per annum. As he has a good income he can afford to repay an appreciable sum. He therefore agrees to make regular monthly payments of £205. Is his mortgage repaid within nine years or not? How much does he owe at the end of that period? (It is assumed that the rate of interest remains constant.)

In this case r is for one month, i.e. $r = \frac{11}{12}$ ($\frac{1}{12}$th of annual interest, 11%).

Likewise n is *not* the number of years but the number of monthly repayments,

i.e. $$n = 12 \times 9 = 108$$

First *month's* interest due is

$$\frac{14\,000 \times 11}{100 \times 12}$$

$$\therefore \qquad E = 205 - \frac{14\,000 \times 11}{100 \times 12}$$

1	4	0	0	0	×	11	÷	1	0	0

÷	1	2	+/−	+	2	0	5	=

76.666667

$$\therefore \quad P_{108} = 14\,000 - \frac{100 \times 76.666667 \times 12}{11}\left\{\left(1 + \frac{11}{1200}\right)^{108} - 1\right\}$$

$$= 14\,000 - 14\,043.586$$

$$= \mathbf{-£43.59}$$

It is obvious that Edwards paid off the loan during the previous month, i.e. after 8 years 11 months. The amount payable in the final month could be calculated from

$$14\,000 - \frac{100 \times 76.666667 \times 12}{11}\left\{\left(1 + \frac{11}{1200}\right)^{107} - 1\right\}$$

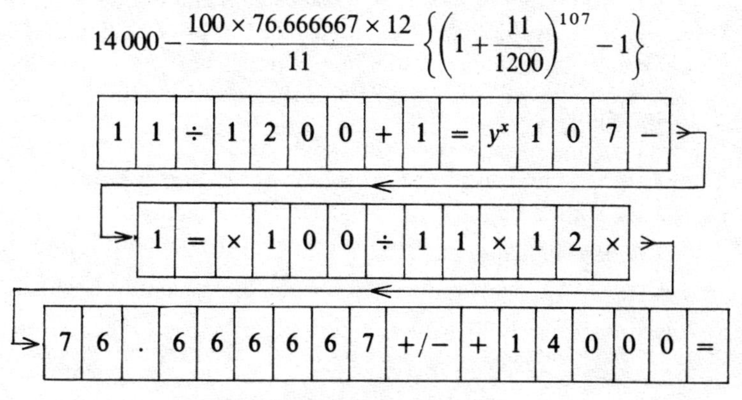

1	1	÷	1	2	0	0	+	1	=	y^x	1	0	7	−

1	=	×	1	0	0	÷	1	1	×	1	2	×

7	6	.	6	6	6	6	6	7	+/−	+	1	4	0	0	0	=

159.94872 ≃ **£159.95**

∴ Edwards owes nothing after paying *106 instalments* of £205 and *one instalment* of £159.95.

Please remember that on each occasion when there is a change in the interest rate charged, or a modification made to the regular instalment paid by the borrower, a new calculation must be made as from that day.

EXERCISE 5

N.B. In questions 2 and 3, interests are assumed to be net (after relief of tax).

1 Shaw borrows £2000 at 14% per annum interest. He repays £500 at end of each of the first five years. How much capital does he still owe immediately after the fifth payment?

2 Smith requires a mortgage of £18 000 on which interest is to be charged at 12% per annum. He agrees to pay regular monthly instalments. Find how much capital he still owes at the end of five years, if his monthly instalments are (a) £180, (b) £240, (c) £300.

3 Jones takes out a mortgage of £16 800 at $13\frac{1}{2}$% p.a., repayments being made on a monthly basis. He decides that he can afford to make these payments £225. After 4 years 5 months the mortgage rate is reduced to $10\frac{1}{2}$%. How much does he still owe in capital at the end of eight years? (*Hint:* Calculate the mortgage outstanding after 4 years 5 months, i.e. 53 months; then start again, using the new interest rate for the remaining period of time, i.e. 3 years 7 months.)

4 Thompson requires a loan of £2500 to help him to buy a car. He agrees to repay this loan by yearly instalments of £600. If interest is charged at 15% on the outstanding loan each year, how much does he have to pay to settle the debt at the end of the seventh year? (This will include a modest excess over £ 600.)

7 Use of [log] key

Earlier in this chapter a logarithmic key was used, without comment, in place of [y^x], in case the latter did not appear on the keyboard.

It is, in fact, rarely that the need arises for [log], when dealing with the sort of problem investigated in this book, but there is one application which could be particularly helpful. It occurs when one wishes to know, for example, in how many years a sum of money invested at compound interest will double itself, or will reach some specified amount.

To find the logarithm of a number, say 7.5, we enter | 7 | · | 5 | log | ; the 'equals' sign is not required. The result is 0.87506126. To find an antilogarithm we merely use | INV | log | ; e.g. antilog 1.2, | 1 | . | 2 | INV | log | 15.848932.

Example 13 In how many years does a sum of money double itself, if left invested at 9.5% tax-free interest payable annually?

Let the sum invested be £P, then it needs to become £$2P$. Take the number of years to be n, then

$$£P \times (1.095)^n = £2\,P, \text{ i.e. } (1.095)^n = 2$$

The magnitude of the sum invested is immaterial, hence, taking logarithms

$$n \log (1.095) = \log 2 \text{ i.e. } n = \log 2 \div \log 1.095.$$

From the calculator,

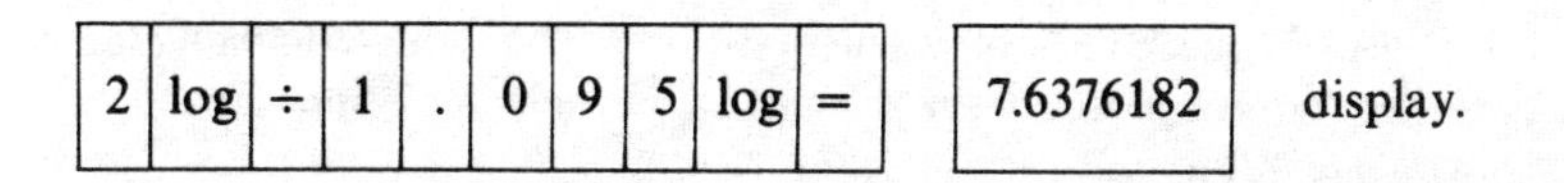

The time required is a little under **7 years 8 months.**

Those readers who wish to study more about the elementary theory of logarithms are referred to *Arithmetic* (Teach Yourself Books), Chapters 12 and 14.

11

Trigonometry and the Electronic Calculator

1 The meaning and origin of trigonometry

Trigonometry is the branch of mathematics which is concerned with relationships which exist between the sides and angles of triangles, and their measurement. It includes a study of trigonometric functions and their uses in 'solving' triangles, i.e. in finding unknown lengths, angles and areas, when adequate information is provided. The subject is applied, among other things, to building projects, surveying, civil engineering, navigation and astronomy. It is also a powerful theoretical tool in the hands of mathematicians. The name *trigonometry* is derived from the Greek: *trigōnon* = a triangle, *metron* = measure.

It has been conjectured that the ancient Egyptians had some rudimentary ideas of the subject and applied them, for example, in the building of the pyramids. The evidence is inconclusive and it is widely accepted that the true foundation of trigonometry was laid by Hipparchus (*c.* 160 BC–*c.* 125 BC), a renowned Greek astronomer. Among other achievements, he calculated the length of the solar year, catalogued at least 800 stars, and advanced the contemporary knowledge of latitude and longitude. His work is included in a masterpiece of the second century AD, namely the *Almagest* (*c.* 150 AD) of Ptolemy of Alexandria (*c.* 100 AD–*c.* 178 AD), the great Egyptian astronomer.

Trigonometry comes in two forms, plane and spherical. The latter is a study of curved triangles lying on the surface of a hypothetical sphere; it is interesting, especially to astronomers and navigators, but it is not suited to this book, in which we shall restrict ourselves to the former, plane, variety.

The present chapter is a shortened course of trigonometry, specifically geared to modern high-speed solution by electronic calculator; tables are not used – they are, in fact, unnecessary.

We start with some properties of similar triangles.

2 Similar triangles

Definition Triangles which have corresponding angles equal are said to be *similar*. They are of the same shape but not of the same size. (Had they been of the same shape and size, they would have been congruent.)

THEOREM 9 If two triangles are similar, their corresponding sides are proportional (i.e. are in the same ratio). The converse is also true: if the corresponding sides of two triangles are proportional, the triangles are equiangular (i.e. corresponding angles are equal), hence they are similar, by the above definition.

We now need three notes:

(i) Corresponding sides are opposite corresponding angles,

(ii) The symbol ||| means 'similar to',

(iii) The abbreviation (AAA) means that two triangles are similar because their corresponding angles are equal. In fact, we only need to know that two corresponding angles are equal. This is because the three angles of a triangle always add up to 180°. Hence (AA) is sometimes used instead of (AAA).

In Fig. 69 below, $\triangle$s ABC and DEF are equiangular:
$\hat{A} = \hat{D}$, $\hat{B} = \hat{E}$, $\hat{C} = \hat{F}$, and hence $\triangle$ABC ||| $\triangle$DEF (AAA).

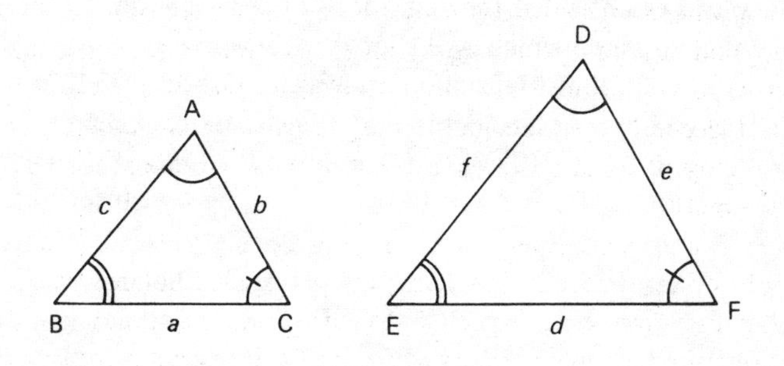

Fig. 69

hence $$\frac{a}{d} = \frac{b}{e} = \frac{e}{f}$$

N.B. Corresponding order *must* be strictly maintained in the presentation of vertices, angles and sides of similar triangles.

Example 1 The given triangles are similar. Find x and y.

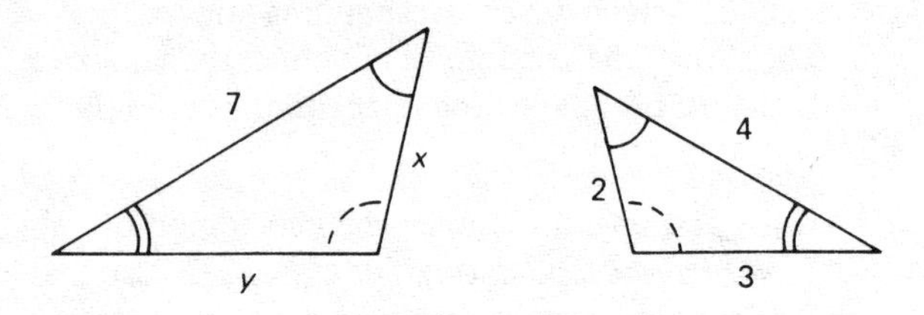

The corresponding ratios are

$$\frac{x}{2} = \frac{y}{3} = \frac{7}{4} \qquad \text{(taking sides opposite equal angles shown)}$$

$$\therefore \quad \frac{x}{2} = \frac{7}{4} \Rightarrow x = \frac{7 \times 2}{4} = \mathbf{3.5}; \; \frac{y}{3} = \frac{7}{4} \Rightarrow y = \frac{3 \times 7}{4} = \mathbf{5.25}$$

The case of *superimposed* similar triangles, when the triangles have one *common* angle, i.e. two of the sides of the smaller triangle lie along the two corresponding sides of the larger triangle, is of great importance in trigonometry when the triangles are right-angled. The fact that they are of this specific type does not affect the theorem above, but is essential for the definitions of *trigonometric ratios* (sine, cosine, tangent, etc.) which will shortly be given.

Let us now look at two similar triangles PQR and PST, in which angle P is common (coincident), and PQ lies along PS and PR lies along PT (Fig. 70). In △ PST, Q is a point on PS; QR is drawn parallel to ST, meeting it at T. In △ s PQR, PST, using an earlier notation

$$y_1 = y_2 \text{ and } z_1 = z_2 \quad \text{(QR} \parallel \text{ST; corresponding angles)}$$

$$\therefore \qquad \triangle \text{PQR} \,|||\, \triangle \text{PST} \qquad \text{(AA)}$$

hence $$\frac{\text{PQ}}{\text{PS}} = \frac{\text{PR}}{\text{PT}} = \frac{\text{QR}}{\text{ST}}$$

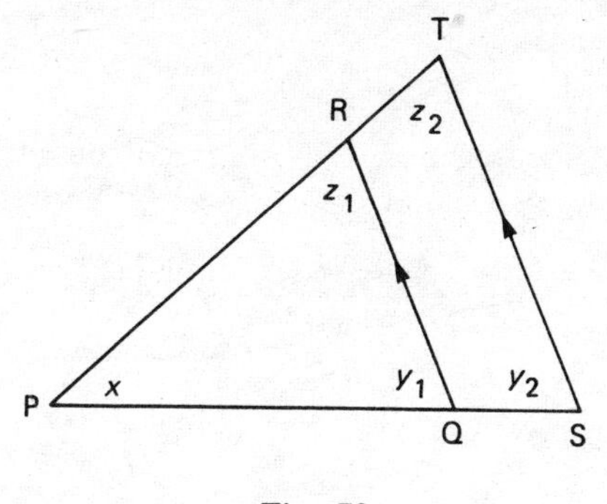

Fig. 70

but note that the ratios can be inverted *together*; this follows from reversing the order of the triangles, because $\triangle PQR \,|||\, \triangle PST \Rightarrow \triangle PST \,|||\, \triangle PQR$. Hence

$$\frac{PS}{PQ} = \frac{PT}{PR} = \frac{ST}{QR}$$

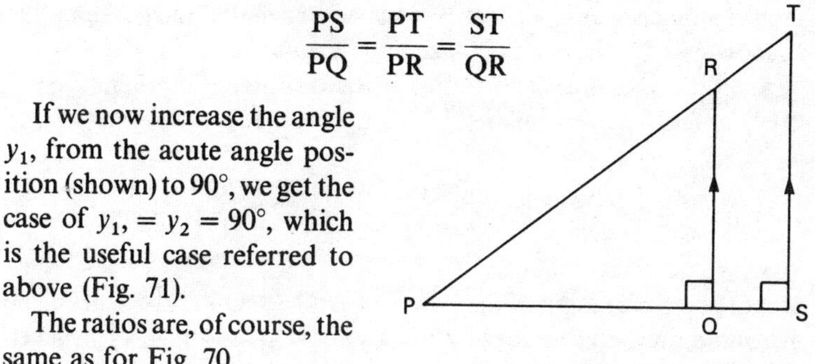

If we now increase the angle y_1, from the acute angle position (shown) to 90°, we get the case of $y_1, = y_2 = 90°$, which is the useful case referred to above (Fig. 71).

The ratios are, of course, the same as for Fig. 70.

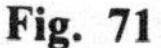

Fig. 71

Problem To find the height of a square pyramid, as measured by Thales of Miletus (*c.* 640 – 546 BC), a highly successful merchant who later turned his attention to mathematical and scientific reasoning. He devised the idea of shadow reckoning and he used similar triangles in the process (Fig. 72). It is quite possible that the work of Thales may have helped Hipparchus to establish his basis of trigonometry.

The method of solution of the problem was ingenious, and was very probably akin to the following procedure. It is worthy of study, even in these days of far more sophisticated mathematics.

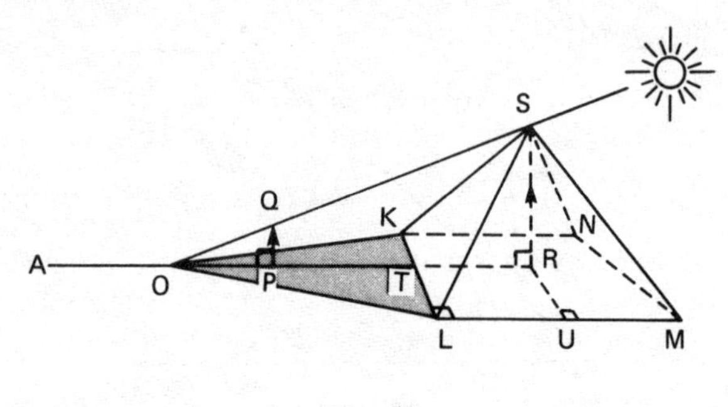

Fig. 72

The pyramid has vertex S and square base KLMN; R, the centre of the square, lies vertically below S.

The line AT is laid out, on the ground, so that it perpendicularly bisects the edge KL. One waits until the shadow KOL, cast by the sun, has its vertex at O, *somewhere on AT*. A pole is then held vertically and moved along AT until a point P is reached, such that the (straight line) shadow cast by the pole (PQ) exactly terminates at O.

As PQ and RS are vertical, and OPR is a straight line, then triangles OPQ and ORS are similar ($\hat{O}$ is common; $O\hat{P}Q = 90^\circ = O\hat{R}S$).

$$\text{Hence } \frac{RS}{OR} = \frac{PQ}{OP} \Rightarrow RS = \frac{PQ \times OR}{OP} = \frac{PQ \times (OT + LU)}{OP}$$

where PQ is the length of the pole; $LU = \frac{1}{2}LM$, OT and OP are easily measured on the ground (which is, by the way, assumed to be level).

EXERCISE 1

1 In $\triangle$ ABC, AB = 5 cm, AC = 7 cm and BC = 10 cm. In $\triangle$ DEF, EF = 8 cm. If $\triangle$ ABC ||| $\triangle$ DEF, find the lengths of DE and EF.

2 In $\triangle$ABC, AB = 12 cm and AC = 15 cm. D is a point on AB such that AD = 8 cm. Draw DE || AC, meeting BC at E. (i) Prove that $\triangle$ BDE ||| $\triangle$ BAC. (ii) Calculate the length of DE.

3 The sides of a triangle are of length 4, 5 and 7 cm respectively. Construct a similar triangle which has its longest side of length 6 cm. What are the lengths of the other sides? (Refer to page 104.)

3 The tangent of an angle

In the work above, it will have been noticed that, given three measurements, we can find a fourth one by calculation, e.g. given $a = 2$, $b = 3$, $c = 4$, then if $\frac{a}{b} = \frac{c}{d}$, we have $d = 6$. Can we, however, reduce the number of measurements of lengths and/or angles needed, in a triangle, in order to find any or all of the remaining ones? Indeed we can, by means of trigonometry, exploiting the properties of right-angled triangles.

Before we go on, it must be made clear to those unfamiliar with trigonometry that the meaning of *tangent* in this chapter is not the geometrical idea of a straight line which touches a circle (or some other curve); the present significance of tangent is that of being a particular kind of ratio.

Consider an angle A, having two arms, AB and AC. Take points P, Q, R, S . . . on AB, and from them draw perpendiculars to AC meeting it at K, L, M, N . . . respectively (Fig. 73).

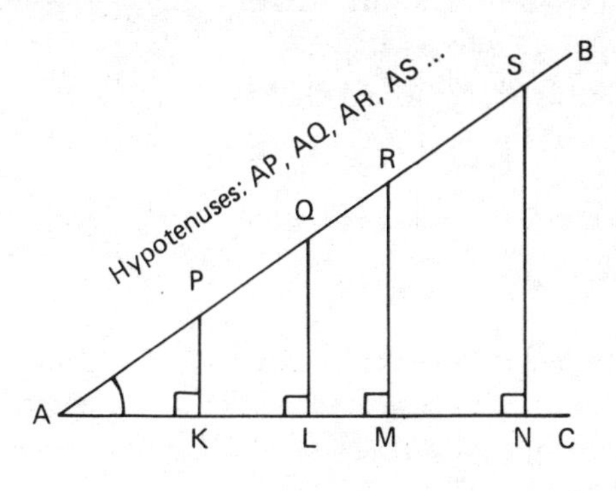

Fig. 73

Then in $\triangle$ s AKP, ALQ, AMR, ANS . . .

we have $\quad A\hat{K}P = A\hat{L}Q = A\hat{M}R = A\hat{N}S$ . . . (*all* right angles)

and $\quad \hat{A}$ is common to all the triangles

$\therefore \quad \triangle AKP \;|||\; \triangle ALQ \;|||\; \triangle AMR \;|||\; \triangle ANS \ldots \qquad$ (AA)

$$\therefore \quad \frac{KP}{AP} = \frac{LQ}{AL} = \frac{MR}{AM} = \frac{NS}{AN} \ldots$$

In fact, for a given angle A, if we take any point on AB and draw a perpendicular to AC, the ratio $\dfrac{\text{Opposite side (to angle A)}}{\text{Adjacent side (to angle A)}}$ is *constant.*

No confusion arises with regard to the third side of any right-angled triangle in Fig. 73, for in every case it is the hypotenuse (the side opposite the right angle).

Definition In a right-angled triangle, the *tangent* of an angle is the ratio of its opposite side to its adjacent side. The tangent of angle A, say, is written as tan A, and it is exactly the ratio given above. In Fig. 73, for example:

$$\tan A = \boxed{\frac{KP}{AP} = \frac{LQ}{AL} = \ldots} = \frac{\text{opposite side}}{\text{adjacent side}}$$

There are various memory aids to help one to recall the different trigonometric ratios. My favourite one, for tangent, suggested by a student some twenty years ago, is Tanks Oppose Artillery (*t*angent = *o*pposite ÷ *a*djacent). Other mnemonics for different trigonometric ratios follow later.

Look at Fig. 74, showing $\triangle ABC$ with $\hat{C} = 90°$.

We have

$$\tan A = \frac{a}{b} \Rightarrow b \tan A = a, \text{ i.e. } a = b \tan A$$

Likewise

$$\tan B = \frac{b}{a} \Rightarrow a \tan B = b, \text{ i.e. } b = a \tan B$$

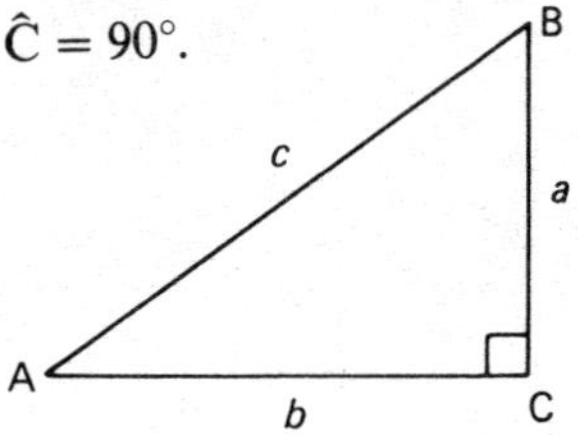

Fig. 74

where, as $\hat{C} = 90°$, then $\hat{A} + \hat{B} = 90°$, i.e. $\hat{A} = 90° - B$ and $\hat{B} = 90° - A$

$\hat{A}$ and $\hat{B}$ are in fact *complementary*.

Note: It will probably have already been noticed that the opposite side of angle A is the adjacent side of angle B; likewise, the adjacent side of angle A is the opposite side of angle B.

4 The graph of tan A from 0° to 90°

Before we begin to use a calculator, which is extremely fast but does not show us *why* we get certain results, it is desirable to understand something of the characteristic of trigonometric functions. We therefore commence by constructing a graph of the tangent of an angle, in the range 0° to 90°, using basic instruments only: pencil, ruler, protractor (and possibly set square) and graph paper.

Consider $\triangle$ ABC, right-angled at C, and take AC = 1 unit. This may be done without loss of generality, as we already know (Section 3) that altering the *size* of a triangle while retaining its shape does not affect the ratio tan A = CB ÷ AC.

Now AC = 1, $\therefore$ tan A = CB ÷ 1 = 1 (unit)

In order to sketch the graph, we take a base-line AB = 1 unit. (An actual length of 2 cm is recommended.) We erect CB $\perp$ AC. With a protractor, we lay off angles $CAB_0 = 0°$, $CAB_1 = 15°$, $CAB_2 = 30°$ . . . $CAB_5 = 75°$, meeting CB at B_0 (which coincides with C), B_1, B_2 . . . B_5, respectively. We cannot, of course, use $CAB_6 = 90°$, for this would be parallel to CB, and would (as we say mathematically) meet it at infinity. The values of CB_1, CB_2, . . . CB_5 are plotted against the corresponding values of angle A, namely, 15°, 30° . . . 75° (Fig. 75). *The reader is recommended to do this for himself.*

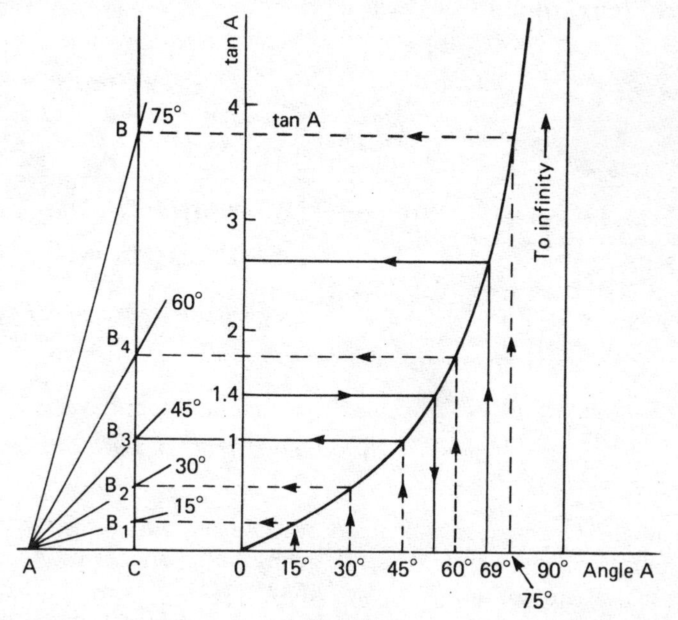

Fig. 75

We are now ready to read off approximate values of tan A from the graph, for *any* angle from 0° to 75°, or a little beyond this. (If we go much further, the values of the tangents rapidly increase and do not conveniently fit on to graph paper, e.g. tan 88° $\simeq$ 28.64, tan 89° 30′ $\simeq$ 114.59, and tan 90° is infinite, ∞.) Conversely from the graph, if we know the tangent of an angle, we can find the angle itself. (See Example 3 below).

Firstly, though, it is time to look at additional keys on our calculator which, in addition to those we have already used, is now assumed to have trigonometric keys marked [SIN], [COS] and [TAN]. The last-named is given below, and the other two follow shortly.

If one refers back to page 53 (Section 3), it will be recalled that 1° is subdivided into 60′ (minutes of arc). In using an angle such as 27° 49′ it is best to convert to decimal form firstly on a calculator.*

Example 2 Using a calculator, find the values of tan 2°, tan 36.7° and tan 69° 31′. Give the results (i) in full, (ii) correct to four decimal places, (iii) correct to two decimal places.

Calculator keys used	To accuracy of calculator	4 dec. pl.	2 dec. pl.
2 TAN →	0.03492077	0.0349	0.03
3 6 . 7 TAN →	0.74537703	0.7454	0.75
* 3 1 ÷ 6 0 + 6 9 = TAN →	2.6769951	2.6770	2.68

It is *not* necessary to press [=] in order to get the *tangent* evaluated.

Example 3 Let us now compare results from our calculator with results from our graph (Fig. 75), giving both sets correct to two decimal places, for angles 0°, 15°, 30°, 45°, 60°, 75°.

From the graph, find (i) tan 90°, (ii) the angle whose tangent is 1.4.

ANGLE	TANGENT (CALCULATOR	TANGENT (GRAPH)
0°	0.00	0.00
15°	0.27	0.27
30°	0.58	0.58
45°	1.00	1.00
60°	1.73	1.74
75°	3.73	3.72

{Why is it exactly 1?

(i) From our graph, tan 69° $\simeq$ **2.60**

(ii) Notice that we reverse our directions of travel, starting at 1.4 on the tan A axis and finishing on the angle A axis: the angle whose tangent is 1.4 is approximately **54**°, i.e. tan 54° $\simeq$ 1.4.

To find the value of this angle *by calculator*, we now require another key, [INV]. This key gives the *inverses* of certain of the functions on the machine, e.g. [sin], [cos], [tan], [y^x], [ln x] [log x], some of which have not yet been met in this book. The [INV] key must *not* be confused with [$1/x$], which merely gives the *reciprocal* of a number, i.e. one divided by the number. (The reciprocal of x is $\frac{1}{x}$.)

Definitions

The *angle of elevation* is the *angle* between the horizontal plane and an oblique line from an observer's eye to a specific point above his eye level.

The *angle of depression* is the angle between the horizontal plane and an oblique line from an observer's eye to a specific point below the level of his eye.

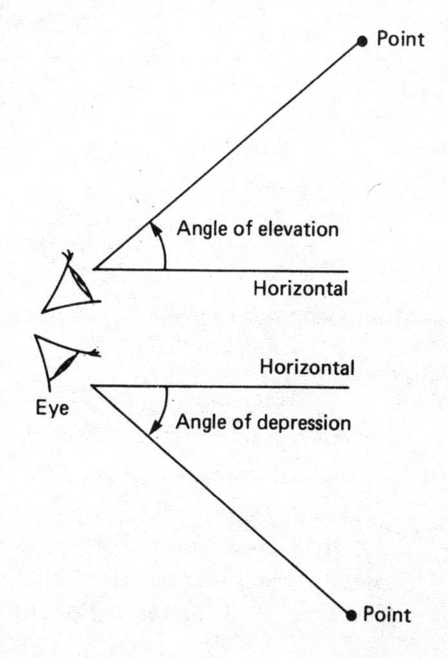

Example 4 The angle of elevation of the top of the Eiffel Tower is found to be 58° 12′ by an observer who is standing on level ground at a distance of 186 m from the central base point of the tower. Find the height of the Tower, to the nearest metre.

Although it is not necessary to draw the Tower in detail, it is instructive to do so. If, for instance, we had used the edge of one foot, we would have had an obtuse-angled triangle (△ EFT); the one we *chose* was right-angled (△ ECT), which is easier to use.

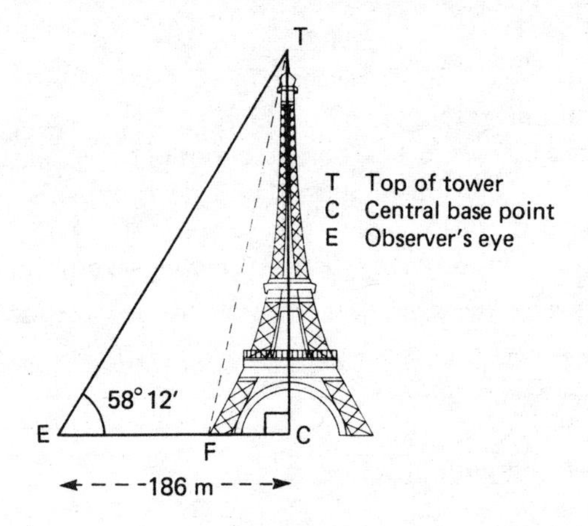

Fig. 76

From △ ECT in Fig. 76,

$$\frac{CT}{EC} = \tan 58° 12'$$

Multiplying both sides by EC,

$$CT = EC \tan 58° 12' = 186 \tan 58° 12'$$

1	2	÷	6	0	+	5	8	=	TAN	×	1	8	6	=

$$= 299.98\ldots \text{ m} \simeq \mathbf{300\ m}$$

Notes: (i) The height is correct, but excludes the television antennae added long after the tower was built. (ii) The observer's eye was assumed to be placed at ground level, but we normally take his height of eye into account. It used to be taken frequently as 5 ft, but now we are metricated 1.5 m is probably near enough.

Suppose we allow for this 1.5 m. What would the correct angle of elevation be?

We would now have

$$186 \tan E = 300 - 1.5 = 298.5$$

$$\therefore \qquad \tan E = \frac{298.5}{186}$$

hence $$\hat{E} = \textit{inverse tan}\ (298.5 \div 186) \quad \ldots\ldots (1)$$

i.e. the angle $\hat{E}$ is the angle whose tangent has the value of $298.5 \div 186$ (which happens to be 1.6048387).

Returning to equation (1) above and doing the job properly by calculator, we have

2	9	8	.	5	÷	1	8	6	=	INV	TAN

= **58.**0723°

.	0	7	2	3	×	6	0

= 4.339′ ≏ **4′**

∴ the required angle is **58° 04′** (we insert the 0 to keep the unit 4 in its correct position.)

This makes a nice example of the use of the [INV] key.

EXERCISE 2

1 Use your calculator to find the values of the tangent of each of the following angles, giving the result correct to 4 decimal places where necessary: (a) 5°, (b) 29°, (c) 46°, (d) 78°, (e) 14° 30′, (f) 33°06′, (g) 47° 54′, (h) 69° 13′, (i) 15°08′, (j) 56° 51′, (k) 21°48′, (l) 63° 26′, (m) 0°05′, (n) 89° 30′.

2 Find, to the nearest 1′ where necessary, the acute angles whose tangents are as follows, but remember, for example, that 39.30° is *not* 39° 30′: (a) 0.7003, (b) 6.3138, (c) 0.2736, (d) 1.7045, (e) 0.9965, (f) 0.1453, (g) 0.6813, (h) 1.1504, (i) 4.5473, (j) 86, (k) 1.

3 Calculate the following, giving the results correct to two decimal places: (a) $9 \tan 71°34'$, (b) $16 \tan 21°48'$, (c) $2.5 \tan 52°30'$, (d) $4.3 \tan 84°37'$, (e) $3 \tan 47° - 4 \tan 38°48'$, (f) $67.5 \tan^2 58°$ (find the value of tan 58°, square, then multiply by 67.5), (g) $8.64 (\tan^2 75° - \tan^2 60°)$.

4 ABC is a triangle in which $\tan A = 0.9$ and $\tan B = 1.4$. What is the value of $\hat{C}$, to the nearest 1′?

5 Find the values of x, y and z in the following diagrams, giving the results correct to one decimal place.

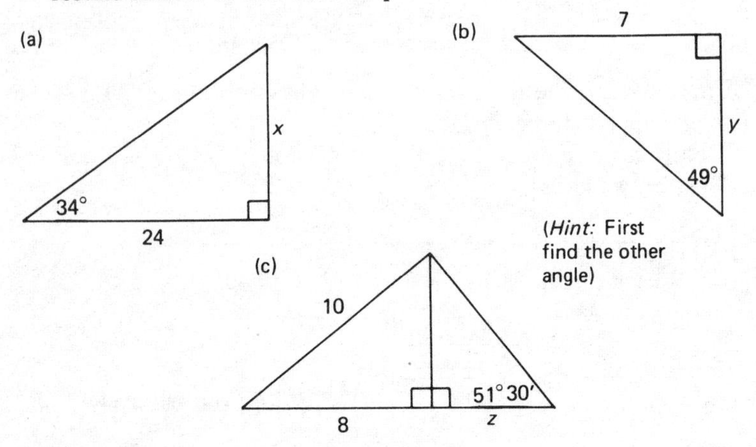

6 A ladder leaning against the wall of a house makes an angle of 76° with the horizontal ground. The top of the ladder is 1.5 m below the top of the wall, and the foot of the ladder is 1.22 m from the wall. Calculate the height of the wall, to the nearest 0.1 m. (Assume that the ladder is in a vertical plane ⊥ to the wall.)

7 △ABC is right-angled at C; AB = 18 cm and BC = 15 cm. Find angle B. (*Hint*: Firstly find the missing side by Pythagoras' theorem. As will be seen later, this question can be answered more quickly by using a different trigonometric ratio.)

8 The diagram represents a cross-section of a ridge-tent on horizontal ground. The base width QR = 1.6 m, and PQ = PR. If angle P = 67°, find the height OP above the ground, to the nearest centimetre.

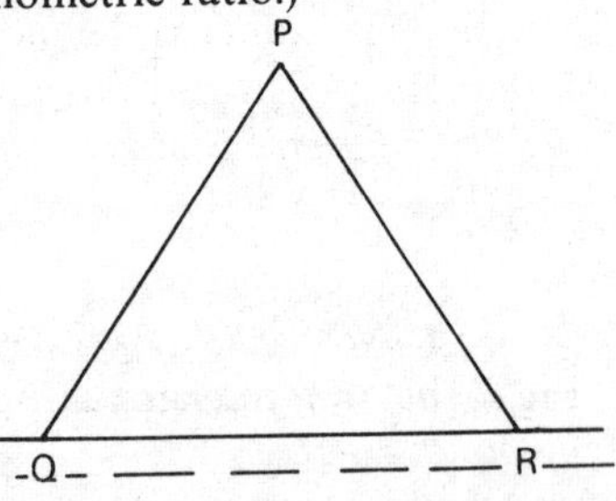

5 More trigonometric functions

It has already been mentioned that posterity owes a considerable debt of gratitude to Ptolemy of Alexandria for including in his *Almagest* the pioneering work of Hipparchus on trigonometry. It was not, however, until much later that our modern names for the trigonometric ratios were created and ultimately became universally accepted. There are in all six *basic* functions:

Names	sine	cosine	tangent	cosecant	secant	cotangent
Written in expressions as	sin	cos	tan	cosec *or* csc	sec	cot

For example, we may find sin A, cos X, tan x, sec (A + B), $\cos^2 3A$, but we shall only deal with straightforward functions.

The tangent we have already seen. Sine and cosine follow in the next section. The other three will not appear in this short summary of trigonometry, save in the following comments:

(i) They are the reciprocals of the first three, namely,

$$\text{cosec}\,X = \frac{1}{\sin X}, \quad \sec X = \frac{1}{\cos X}, \quad \cot X = \frac{1}{\tan X}.$$

(ii) If, at a later date, the reader wishes to find, say, the value of cot 70°, he merely needs to use one extra key, thus [7] [0] [TAN] [1/x].

The names developed in the same haphazard way as did the arithmetical signs, +, −, etc. One man invented, in Latinised form, *sinus* (12th century), which had erratic designations until *sine* (sin) became standard in the seventeenth century. Our ratio *cosine*, first became approximated thereto by Gunter who called it *cosinus* (early 17th century). The name *tangent* is ascribed to Vieta (see Chapter 1) in the late sixteenth century.

6 Gemini – sine and cosine

Fig. 77 below is the same as Fig. 74, used earlier when the tangent of an angle was defined.

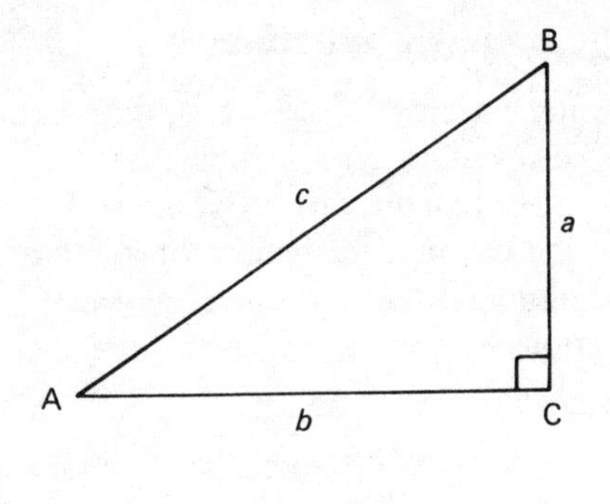

Fig. 77

The sine of an angle

Definition In a right-angled triangle, the *sine* of an angle is the ratio of the opposite side to the hypotenuse.

In Fig. 77, the sine of angle A is written sin A, then

$$\sin A = \frac{\text{opposite}}{\text{hypotenuse}} = \frac{BC}{AB} = \frac{a}{c}; \text{ similarly, } \sin B = \frac{AC}{AB} = \frac{b}{c}.$$

A suitable mnemonic is Shave Our Heads (*s*ine = *o*pposite ÷ *h*ypotenuse).

The cosine of an angle

Definition In a right-angled triangle, the *cosine* of an angle is the ratio of the adjacent side to the hypotenuse.

In Fig. 77, the cosine of angle A is written cos A, then

$$\cos A = \frac{\text{adjacent}}{\text{hypotenuse}} = \frac{AC}{AB} = \frac{b}{c}; \text{ similarly, } \cos B = \frac{BC}{AB} = \frac{a}{c}.$$

A corresponding mnemonic is Coals Are Hot (*c*osine = *a*djacent ÷ *h*ypotenuse).

The sine *increases* (goes up) continuously from 0 to 1, although not at the same rate, when the angle increases from 0° to 90°. The cosine *decreases* (goes down) continuously from 1 to 0, with the same proviso, when the angle increases from 0° to 90°. Unfortunately, for obtuse and reflex angles, these trigonometric ratios undergo changes of direction!

By the way, the abbreviation sin (as in sin X) is pronounced *sīn* (sine) not *sin* (as in original sin), and the corresponding cos (as in cos Y, say) is pronounced *coz* (not *cos*, as in a variety of lettuce).

In order to master elementary trigonometry, the reader, if not already familiar with sine, cosine and tangent, needs to learn these ratios thoroughly. With mnemonics, it is not difficult.

7 Some relationships among sine, cosine and tangent

In Section 6 above, we have seen that, in a right-angled triangle ABC,

$$\left.\begin{aligned} &\sin A = \frac{a}{c} \quad \text{and} \quad \cos B = \frac{a}{c} \quad \therefore \sin A = \cos B, \\ \text{likewise} \quad &\cos A = \frac{b}{c} \quad \text{and} \quad \sin B = \frac{b}{c} \quad \therefore \cos A = \sin B \end{aligned}\right\}$$

but in Section 3 above, using Fig. 74, it was shown that $\hat{A} = 90° - \hat{B}$ and $\hat{B} = 90° - \hat{A}$. Hence, in a right-angled triangle ABC,

$$\sin A = \cos(90° - A). \quad . \quad . \quad . \quad . \quad . \quad . \quad . \quad . \quad (1)$$

and

$$\cos A = \sin(90° - A). \quad . \quad . \quad . \quad . \quad . \quad . \quad . \quad . \quad (2)$$

i.e. the sine of an angle is the cosine of its complement, and *vice versa*.

Returning now to Fig. 77 (the same as Fig. 74), as $\hat{C} = 90°$, in $\triangle$ABC, by Pythagoras' theorem, $a^2 + b^2 = c^2$.

Dividing both sides by c^2,

$$\frac{a^2}{c^2} + \frac{b^2}{c^2} = 1 \Rightarrow \left(\frac{a}{c}\right)^2 + \left(\frac{b}{c}\right)^2 = 1$$

But

$$\frac{a}{c} = \sin A \text{ and } \frac{b}{c} = \cos A$$

$\therefore$

$$\sin^2 A + \cos^2 A = 1 \quad . \quad . \quad . \quad . \quad . \quad . \quad . \quad . \quad (3)$$

Note: $\text{Sin}^2 A$ means $\sin A \times \sin A$; similarly $\cos^2 A = \cos A \times \cos A$, etc. We do *not* write $\sin A^2$ which, if used at all, has a different meaning, and we very rarely use $(\sin A)^2$.

Now consider

$$\frac{\sin A}{\cos A} = \frac{a}{c} \div \frac{b}{c} = \frac{a}{\not{c}_1} \times \frac{\not{c}^1}{b} = \frac{a}{b},$$

but

$$\tan A = \frac{a}{b},$$

$\therefore$

$$\tan A = \frac{\sin A}{\cos A} \quad . \quad . \quad . \quad . \quad . \quad . \quad . \quad . \quad (4)$$

Although we have adhered to angle A, the results found above are true for *any* angles, whether or not they are greater than 90°. The general proofs are omitted.

Example 5 Find the values of the following by calculator, giving the results correct to four decimal places, where necessary: (a) sin 30° (b) sin 72° 15′ (c) cos 68° 32′ (d) cos 37° 22′ (e) sin 114° 30′ (f) cos 114° 30′.

We have, using sin and cos keys,

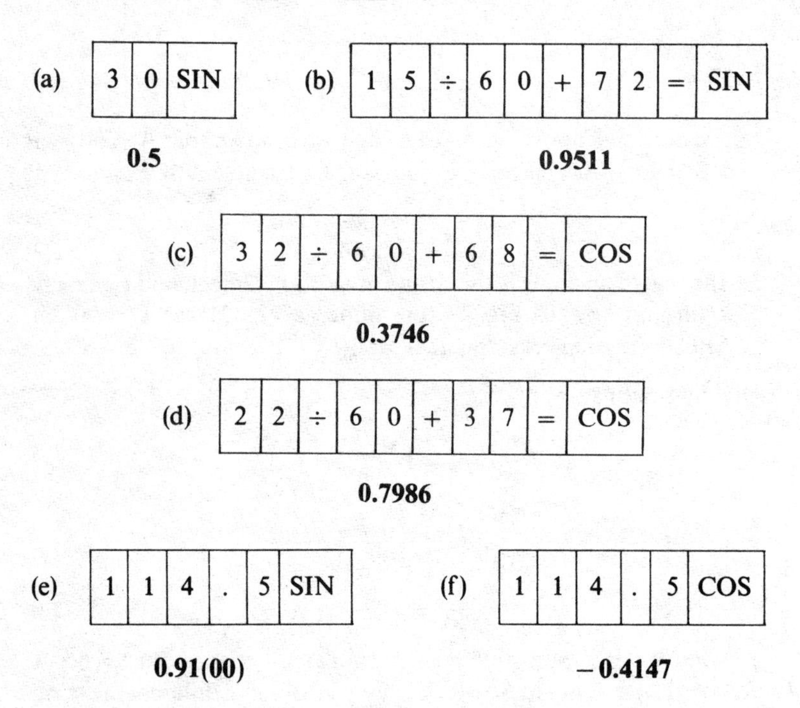

There are some interesting points:

(i) (a) and (b) suggest that sine *increases* from 0° to 90°,
(ii) (c) and (d) suggest that cosine increases from 90° to 0°, i.e. *decreases* from 0° to 90°,
(iii) (e) and (f) suggest that when the angle is *obtuse*, sine is positive, but cosine is negative.

These are purely suggestions, not proofs, but they happen to be correct. Note that 6′, 12′, 18′ . . . 54′ are 0.1°, 0.2°, 0.3° . . . 0.9°; these sometimes save calculator entries. Thus for, say, 47° 18′ we can immediately put [4][7][.][3] instead of [1][8][÷][6][0][+][4][7], a reduction of four key operations.

Example 6 Without quoting the general results just given, show by calculator that (i) sin 37° ÷ cos 37° = tan 37°, (ii) $\sin^2 64° 24′ + \cos^2 64° 24′ = 1$.

(i) We have

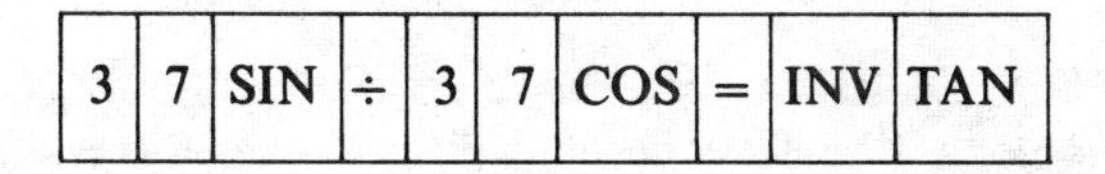

3	7	SIN	÷	3	7	COS	=	INV	TAN

37 (degrees)

After finding the value of sin 37° ÷ cos 37°, we took the inverse of the tangent and found that the result was also 37°.

There is an alternative solution based on a method shown on page 44, though it is not as tidy as the one immediately above. It works as follows:

LHS = [3][7][SIN][÷][3][7][COS][=] 0.753 554 05

RHS = [3][7][TAN] 0.753 554 05

∴ LHS = RHS

(ii) [6][4][.][4][SIN][x^2][+][6][4][.][4][COS][x^2][=] **1**

We used the [x^2] key to square the sine value and then did the same to the cosine value.

Rather more elegant and slightly shorter is the following, but it requires care:

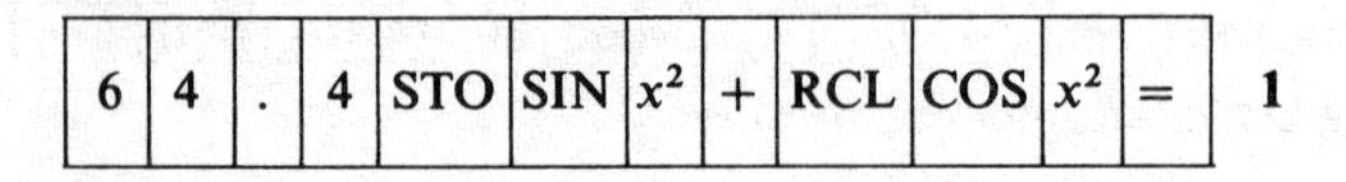

6	4	.	4	STO	SIN	x^2	+	RCL	COS	x^2	=	1

On some calculators, STO may be replaced by M +, and RCL by RM.

EXERCISE 3

1 Use your calculatôr to find the values of the following, giving the results correct to four decimal places where necessary, unless otherwise stated:
(a) sin 45°, (b) $1 \div \sqrt{2}$; what do you conclude from (a) and (b)?
(c) cos 60°, (d) sin 2° 30′, (e) cos 4° 48′, (f) sin 53° 08′, (g) cos 64° 07′, (h) sin 98° 57′, (i) sin 143° 19′, (j) cos 162° 18′

2 Use the keys INV SIN and INV COS (in the same way that INV and TAN were used earlier) to determine, correct to the nearest 1′, the following:
(i) the *acute* angles whose sines are (a) 0, (b) 0.25, (c) 0.59, (d) 0.967, (e) 1, (f) 1.2;
(ii) angles from 0° to 180° whose cosines are (a) 0, (b) 0.1045, (c) 1, (d) 0.6409, (e) −0.472.

3 Compare the values of sin 2Y and 2 sin Y cos Y, when (i) Y = 28°, (ii) Y = 42°.
What do these results suggest? (*Notes*: 2Y merely means 2 × Y as usual; likewise 2 sin Y cos Y is 2 × sin Y × cos Y.)

4 Compare the values of cos 2A and $\cos^2 A - \sin^2 A$, by taking (i) A = 54° 33′, (ii) A = 137° 29′. What do you conclude?

8 The graphs of sine and cosine in the range 0° to 90°

As in Section 5 (Fig. 77), we take triangle ABC which is right-angled at C. We may put the hypotenuse AB = 1 unit without loss of generality, as the size of the triangle does not affect the sine and cosine of an angle, so long as the shape is the same.

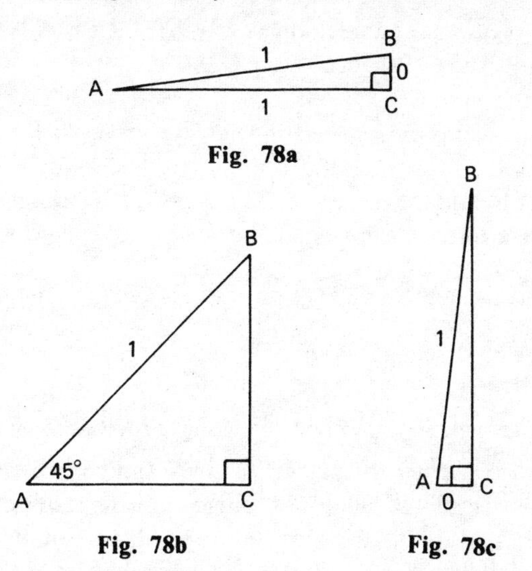

Fig. 78a

Fig. 78b

Fig. 78c

We now look at sin A when A = 0°, 45°, 90° (Figs 78).

In Fig. 78a, we begin with angle A very small, then make A → 0, which gives BC → 0, i.e. when $\hat{A} \rightarrow 0°$

$$\sin A = \frac{BC}{AB} \rightarrow \frac{0}{1} = 0$$

giving $$\sin 0° = 0.$$

In Fig. 78b, as $\hat{A} = 45°$ and $\hat{C} = 90°$,

$\therefore$ $$\hat{B} = 180° - 45° - 90° = 45°$$

i.e. $\hat{A} = \hat{B}$, hence $\triangle$ ABC is isosceles

(Theorem 7 converse)

$\therefore$ $$AC = BC = x, \text{ say} \quad \therefore x^2 + x^2 = 1$$

(Pythagoras' theorem)

i.e. $$x^2 = \frac{1}{2} \text{ giving } x = \frac{1}{\sqrt{2}}$$ (positive root)

$\therefore$ $$\sin 45° = \frac{BC}{AB} = \frac{1}{\sqrt{2}} \div 1 = \frac{1}{\sqrt{2}} \simeq 0.7071$$

(by calculator)

Exercise 3, question 1(a) and (b) above is now solved!

In Fig. 78c, we take angle A as nearly 90°, then as $\hat{A} \to 90°$, we have $AC \to 0$ and $BC \to 1$; hence when $\hat{A} \to 0°$

$$\sin A = \frac{BC}{AB} \to \frac{1}{1} = 1, \text{ giving } \sin 90° = 1.$$

In order to find the corresponding cosines in the range 0° to 90°, we do not have to do all this again, because $\cos A = \sin(90° - A)$,

$$\therefore \text{ when } A = 0° \quad \cos 0° = \sin(90° - 0°) = \sin 90° = 1$$

$$\text{when } A = 45° \quad \cos 45° = \sin(90° - 45°) = \sin 45° = \frac{1}{\sqrt{2}} \simeq 0.7071$$

$$\text{when } A = 90° \quad \cos 90° = \sin(90° - 90°) = \sin 0° = 0.$$

We now construct the graphs of sin A and cos A from 0° to 90°, without the use of a calculator or of trigonometric tables. The method is rather similar to that used for the graph of tan A, in the same range (Section 4 above).

In Fig. 79, we take a horizontal base-line $AC_1 = 1$ unit and we erect $AY \perp AC_1$. With centre A and radius 1 unit, draw a quadrant (quarter of a circle) from AC_1 to AY, cutting AY at B_7. Lay off angles C_1AB_2

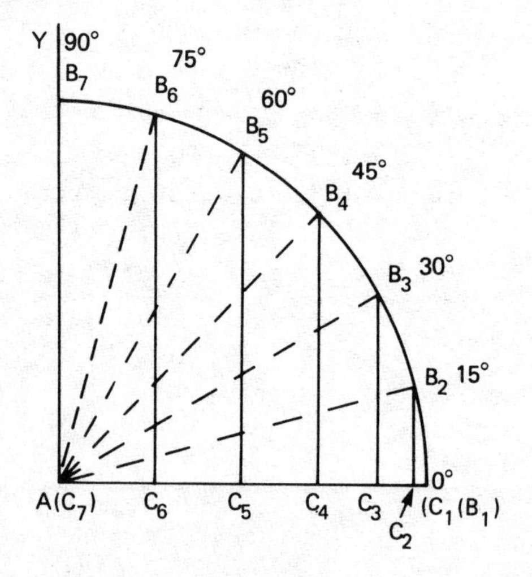

Fig. 79

$= 15°$, $C_1AB_3 = 30°$ . . . $C_1AB_6 = 75°$, at 15° intervals, where all the B's lie on the quadrant. Join AB_2, AB_3 . . . AB_6, and draw B_2C_2, B_3C_2 . . . $B_6C_6 \perp AC_1$, meeting it at C_2, C_3 . . . C_6. The $\triangle$s AB_1C_1, AB_2C_2 . . . AB_7C_7 are thus completed. (*Note*: B_1 coincides with C_1, *and* C_7 coincides with A; hence $\triangle AB_1C_1$ is the degenerate case dealt with in Fig. 78a and $\triangle AB_7C_7$ is the case in Fig. 78c.)

Now, as *every* AB = 1 unit in the diagram,

then $\sin 0° = 0$ (already shown); $\cos 0° = 1$ (already shown);

$$\sin 15° = \frac{B_2C_2}{AB_2} = B_2C_2; \qquad \cos 15° = \frac{AC_2}{AB_2} = AC_2$$

Similarly

$\sin 30°$, . . . $\sin 75°$ give B_3C_3, . . . B_6C_6

$\cos 30°$, . . . $\cos 70°$ give AC_3, . . . AC_6;

and

$\sin 90° = 1$ (already shown)

$\cos 90° = 0$ (already shown).

We enter all the lengths B_1C_1 . . . B_7C_7, *measured* in Fig. 81, in the sine table below, and all the lengths AC_1 . . . AC_7, likewise measured, in the cosine table below

X	Angle (in degrees)	0	15	30	45	60	75	90	Co-ordinates
Y	Sine	0.00	0.26	0.50	0.71	0.87	0.97	1.00	For sine curve

X	Angle (in degrees)	0	15	30	45	60	75	90	Co-ordinates
Y	Cosine	1.00	0.97	0.87	0.71	0.50	0.26	0.00	For cosine curve

We then plot the points whose coordinates are (X, Y) on graph paper, taking OX horizontally and OY vertically upwards (Fig. 80).

We now use the graphs (Fig. 80) to find (i) $\sin 38°$, (ii) angle P, if $\sin P = 0.42$, (iii) $\cos 79°$, (iv) angle Q, if $\cos Q = 0.42$.

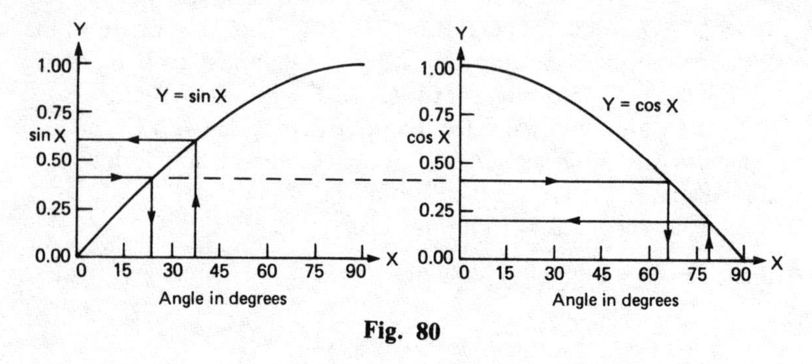

Fig. 80

Our results are (i) $\sin 38° \simeq 0.62$, (ii) angle $P \simeq 25°$, (iii) $\cos 79° \simeq 0.19$, (iv) angle $Q \simeq 65°$. What do you notice about angles P and Q? (The answer is given below, but try it first!)

It will be observed that the cosine curve is a mirror image of the sine curve within the range $0° \leqslant X \leqslant 90°$ (the symbol $\leqslant$ means 'less than or equal to'). This property does *not* necessarily apply to other ranges of angle.

Example 7 A hill of uniform slope rises by a height of 60 m in a distance of 450 m up the slope. What is the angle of inclination of the hill to the horizontal? Express this as a percentage of the angle from horizontal to vertical, to the nearest $\frac{1}{2}$ %.

From the diagram, the required angle is A:

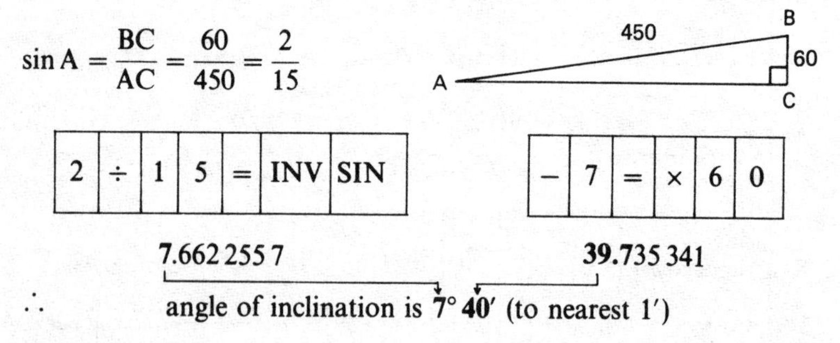

$$\sin A = \frac{BC}{AC} = \frac{60}{450} = \frac{2}{15}$$

2	÷	1	5	=	INV	SIN

−	7	=	×	6	0

7.662 255 7   **39**.735 341

∴ angle of inclination is **7**° **40**′ (to nearest 1′)

For the inclination, to the nearest $\frac{1}{2}$ %, four decimal places are more than adequate. From horizontal to vertical is 90°, hence the percentage is

Answer: $\hat{P} = 25°$ and $\hat{Q} = 65° = 90° - \hat{P}$; but $\sin P = \cos(90° - P)$, from earlier work (Section 6), hence $\sin 25° = \cos 65°$. We have just verified this from the graphs.

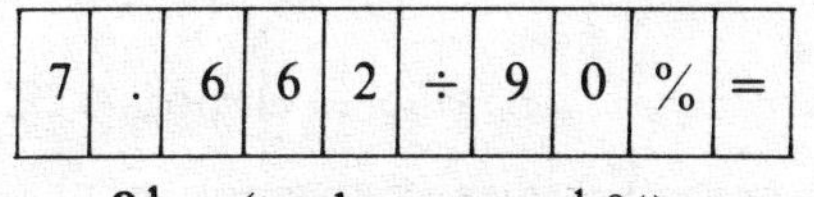

8½ (to the nearest ½ %)

Example 8 In the diagram, PQR is a roof truss, in which $\hat{P} = 56° 23'$, PR = 3 m and QR = 6 m. Calculate the span PQ.

Draw RS ⊥ PQ meeting it at S, then △s PRS and QRS are right-angled at S.

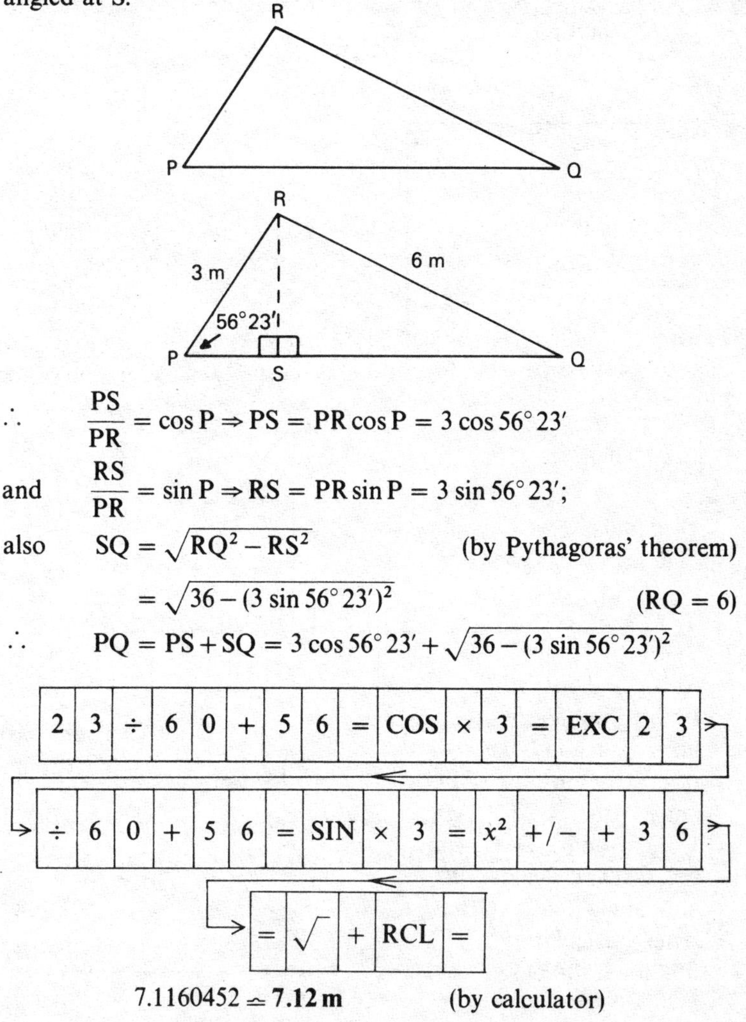

$$\therefore \quad \frac{PS}{PR} = \cos P \Rightarrow PS = PR \cos P = 3 \cos 56° 23'$$

$$\text{and} \quad \frac{RS}{PR} = \sin P \Rightarrow RS = PR \sin P = 3 \sin 56° 23';$$

$$\text{also} \quad SQ = \sqrt{RQ^2 - RS^2} \qquad \text{(by Pythagoras' theorem)}$$

$$= \sqrt{36 - (3 \sin 56° 23')^2} \qquad (RQ = 6)$$

$$\therefore \quad PQ = PS + SQ = 3 \cos 56° 23' + \sqrt{36 - (3 \sin 56° 23')^2}$$

7.1160452 ≃ **7.12 m** (by calculator)

EXERCISE 4

In all these, *except question 5*, use an electronic calculator.

1 Evaluate these expressions, giving the results correct to two decimal places:
(a) $29 \sin 34°$ (b) $17 \cos 76° 12'$ (c) $3.4 \sin 82° 10'$
(d) $4.3 \tan 84° 22'$(e) $\sqrt{3} \sin 30°$ (f) $5 \cos 8° 47'$
(g) $7 \sin 23° \cos 64° 36'$ (h) $6 \tan^2 30°$

2 Find the values of x and y in each of the diagrams shown, correct to two decimal places

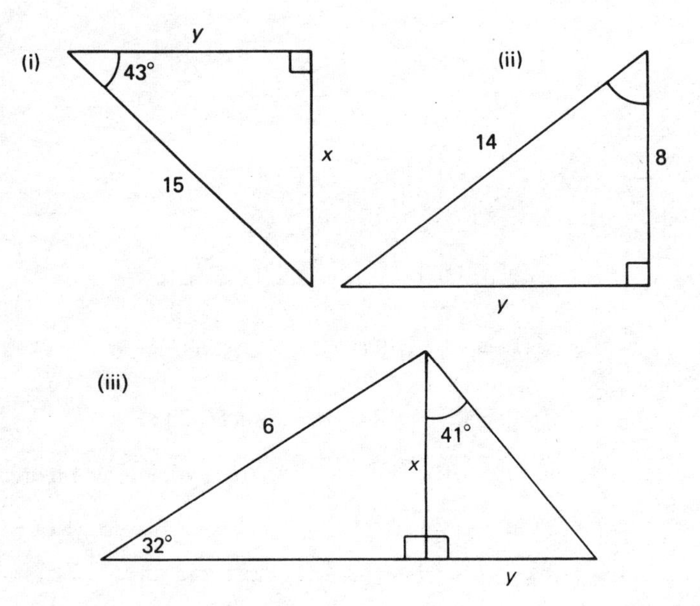

3 In the given figure, AB is horizontal and BD is vertical, C being on BD; angle BAC = 35°, AB = 33 m and CD is 4 m below D. Calculate BC and angle CAD. (*Hint*: After calculating BC, find angle BAD.)

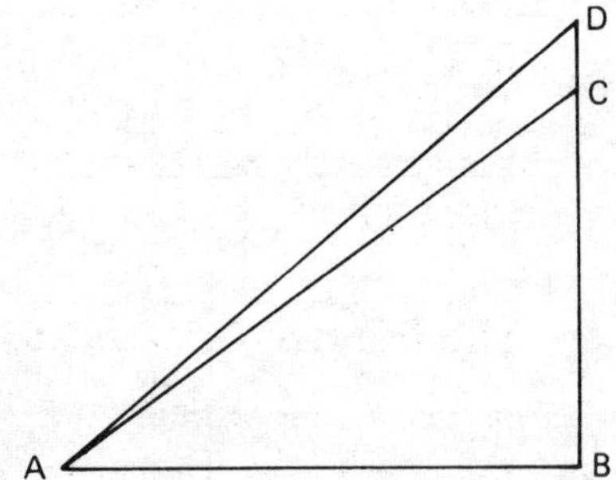

4 In $\triangle XYZ$, angle Y is 90°, XZ = 12.5 cm and YZ = 5 cm. Calculate (i) angle Z, (ii) the length of YW, the perpendicular from Y to XZ.

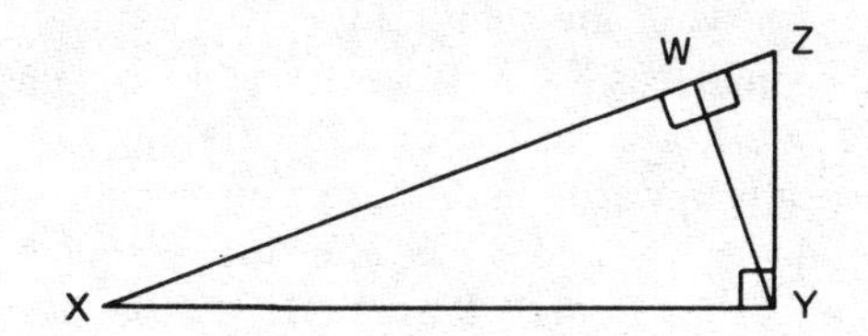

5 In the equilateral triangle ABC, each side is of length 2 units. AM is the perpendicular from A to BC. Prove that $\triangle ABM \equiv \triangle ACM$. How long is BM? Prove that AM = $\sqrt{3}$.

From $\triangle ABM$, find the sine and cosine of (i) 30°, (ii) 60°.

6 Two observers, A and B, standing on level ground, are 25 m apart. They note that the angles of elevation of the top D of a vertical radio mast CD are 31° 44′ and 38° 17′, respectively, ABC being a straight line. Calculate the height of D above ground, adding 1.5 m for the height of eye (assume that the eye levels are the same). (*Hint*: Let BC = x metres and CD = y metres. Find the angles ADC and BDC. Hence get two equations in x, y and tangents of the angles. Substitute for x in terms of y, and solve by calculator.)

9 The trigonometric ratios of obtuse angles

Before starting this section, a brief revision of Chapter 5, Sections 2, 3 and 4, may be helpful. These deal with *directed* numbers, i.e. ones with signs attached, which are important when obtuse angles are involved; furthermore, the facts that, say, $(+p)(+p) = p \times p = p^2$ and $(-p)(-p) = +p^2 = p^2$ are significant.

Draw a horizontal axis X′OX, where O is the origin (0, 0). Numbers measured along OX are positive and those along OX′ are negative. Erect OY vertically upwards; in this direction all numbers are positive. The area bounded by OX and OY is called the *first quadrant* (marked 1 in Fig. 81), and that bounded by YOX′ is the *second* quadrant. (There are two more quadrants, which will shortly receive some coverage.)

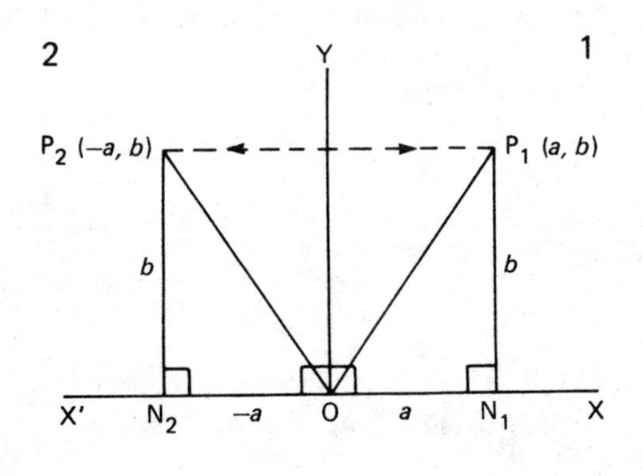

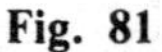
Fig. 81

Let *a*, *b* be two *positive* numbers. We mark the point P_1, whose coordinates are (*a*, *b*) in quadrant 1; then the mirror image of P_1 in the axis OY has coordinates $(-a, b)$. We shall call this point P_2. Join OP_1 and draw $P_1N_1 \perp OX$ meeting it at N_1. Similarly, join OP_2 and draw $P_2N_2 \perp OX'$ meeting it at N_2. We now have $\triangle$s OP_1N_1, OP_2N_2 in Fig. 81. By Pythagoras' theorem:

$$\left.\begin{aligned} OP_1^2 &= ON_1^2 + N_1P_1^2 = a^2 + b^2 \quad \therefore OP_1 = \pm\sqrt{a^2+b^2} \\ OP_2^2 &= ON_2^2 + N_2P_2^2 = (-a)^2 + b^2 = a^2 + b^2 \quad OP_2 = \pm\sqrt{a^2+b^2} \end{aligned}\right\} \text{the same.}$$

By convention we take the *positive* root, hence $OP_1 = OP_2 = +\sqrt{a^2 + b^2}$.

We now extend Fig. 81, as shown in Fig. 82. We draw a semicircle with centre O and radius OP_0 ($= OP_1 = OP_2$), starting at P_0 on OX and finishing at P_3 on OX′. It will, of course, pass through P_1 and P_2. We now look at the *signs* of the sides of the triangles OP_1N_1 and OP_2N_2. They are marked on the diagram.

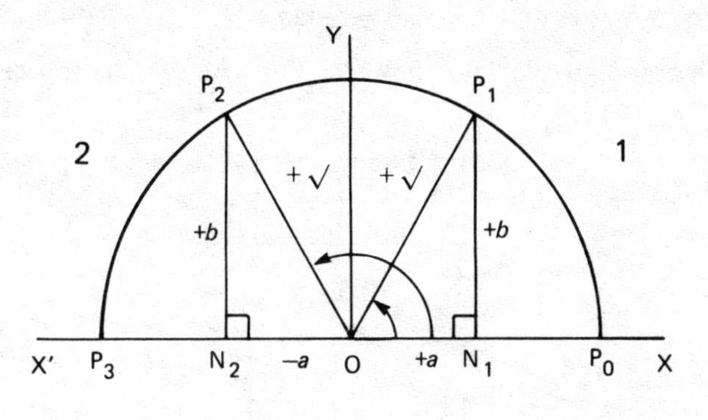

Fig. 82

It is easy to show that $\triangle OP_1N_1 \equiv \triangle OP_2N_2$ (RHS), $\therefore N_1\hat{O}P_1 = N_2\hat{O}P_2$

Let $N_1\hat{O}P_1 = A$, then $N_2\hat{O}P_2 = A$.

Measuring angles *anticlockwise* from OX (the normal method, see Fig. 15), we have

$$N_1\hat{O}P_2 = 180^\circ - N_2\hat{O}P_2 = 180^\circ - A$$

$\therefore$ from the $\triangle$s OP_1N_1 and OP_2N_2

$$\sin A = \frac{b}{\sqrt{a^2+b^2}}, \sin(180^\circ - A) = \frac{b}{\sqrt{a^2+b^2}} \therefore \sin(180^\circ - A) = \sin A$$

$$\cos A = \frac{a}{\sqrt{a^2+b^2}}, \cos(180^\circ - A) = \frac{-a}{\sqrt{a^2+b^2}}$$

$$\therefore \cos(180^\circ - A) = -\cos A$$

$$\tan A = \frac{b}{a}, \tan(180^\circ - A) = \frac{b}{-a} \therefore \tan(180^\circ - A) = -\tan A$$

Example 9 Find the values of (i) $\sin 150°$, (ii) $\cos 128°$, (iii) $\tan 97°$.

We could proceed as above:

(i) $\sin 150° = \sin (180° - 150°) = \sin 30°$
(then use tables or calculator)

(ii) $\cos 128° = \cos (180° - 128°) = -\cos 52°$ (ditto)

(iii) $\tan 97° = \tan (180° - 97°) = -\tan 83°$ (ditto)

With a modern calculator, however, this is likely to be unnecessary, as will be seen:

* (i) $\sin 150°$

1	5	0	SIN

0.5

* (ii) $\cos 128°$

1	2	8	COS

(to four decimal places) **−0.6157**

* (iii) $\tan 97°$

9	7	TAN

(to four decimal places) **−8.1443**

* Any changes of sign in the second quadrant are automatically included in the read-out.

Although we shall not use larger angles, it may be of value to mention that for angles greater than 180°, the correct results will still be obtainable. Here are a few examples in Quadrant 3 (180°→270°) and Quadrant 4 (270° → 360°): see Fig. 83 (page 214).

(iv) $\sin 200° \simeq -0.3420$ (v) $\tan 231° \simeq 1.2349$
(vi) $\cos 342° \simeq 0.9511$ (vii) $\sin 357° \simeq -0.0523$

Again for future use, there follow a few examples of cosecant, secant and tangent. Many calculators do not have keys for these three functions, explicitly. In consequence, the following is a necessary method; remembering that these three ratios are the reciprocals (i.e. $1/x$) of sine, cosine and tangent, respectively.

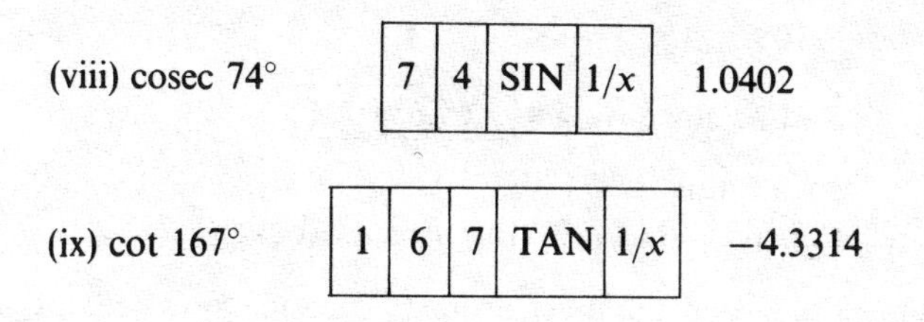

(x) sec 243° 17′

1	7	÷	6	0	+	2	4	3	=	COS	1/x

−2.2243

Example 10 Determine the obtuse angles X, Y and Z, such that (i) sin X = 0.8290, (ii) cos Y = −0.4, (iii) tan Z = −2.4304

(i)

.	8	2	9	INV	SIN

55.9962 ≏ 56° (to nearest 1′)

Now this is acute. The calculator cannot discriminate in this case, because acute *and* obtuse angles have positive *sines*. We therefore have to come to its aid, thus X = 180° − 56° = **124°**

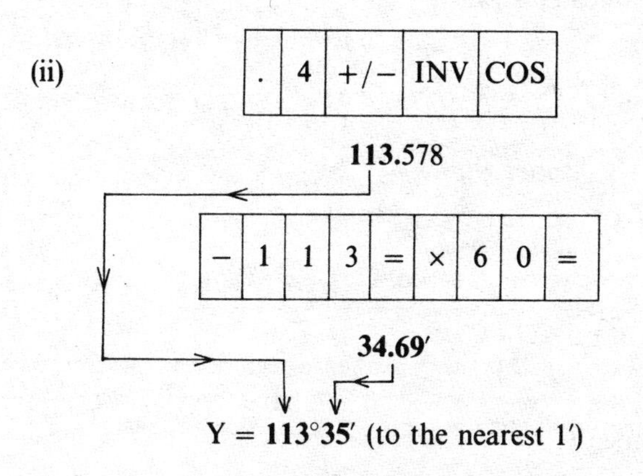

(iii) 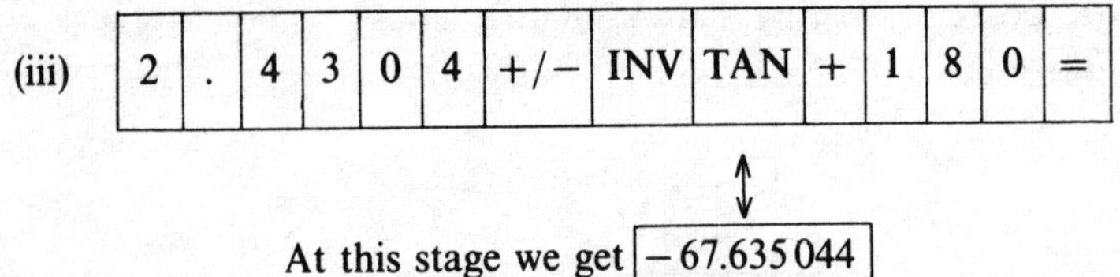

At this stage we get -67.635044

so we must *add* 180°.
112.365 = **112° 22′** (this step as above)

It has just been seen that the inverse sine and cosine keys present little difficulty, whether finding acute or obtuse angles. When dealing with inverse tangent on the calculator used, it is not quite so simple. Example 10 (iii) obtained −68° (roughly), which is in the *fourth* quadrant. Fig. 83 illustrates the point.

By rotating OP_4 through 180° we get the correct direction OP_2, in the second quadrant.

It is 180° − 68° = 112° (to nearest 1°).

It may make the matter clearer if one states the quadrants in which sine, cosine and tangent are *positive*. Where not so stated, they are negative. (As already mentioned, these statements can be proved.)

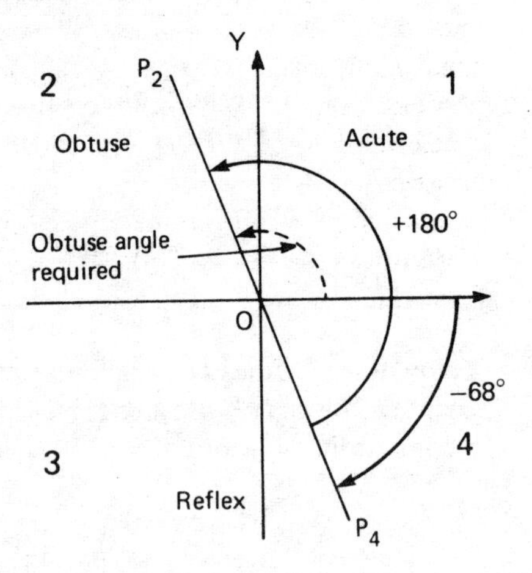

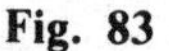
Fig. 83

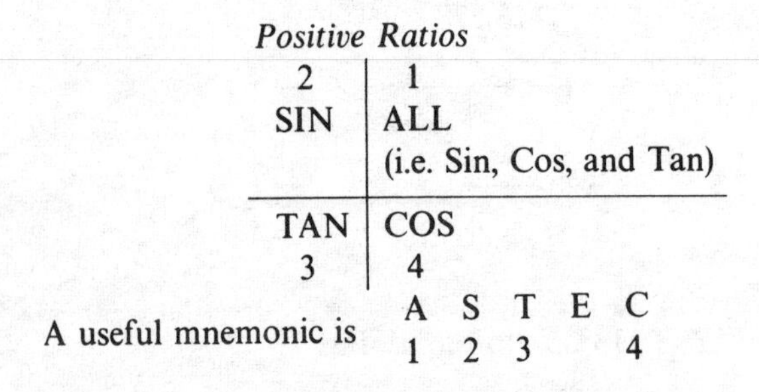

By looking at the negative situation we can see the correlation between the second and fourth quadrant for tangent, in Fig. 83.

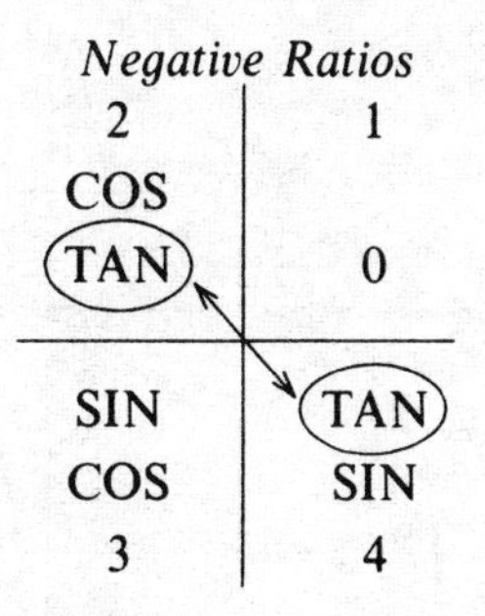

Important note: As the hypotenuse is the longest side of a right-angled triangle, sine, which is *opposite side* ÷ *hypotenuse*, and cosine, which is *adjacent side* ÷ *hypotenuse* cannot be greater than +1 or less than −1. Putting it mathematically, sine and cosine are bounded functions, i.e. $-1 \leqslant \sin A \leqslant 1$ and $-1 \leqslant \cos A \leqslant 1$, for all real values of A.

The tangent is not bounded. It can take values from $-\infty$ to $+\infty$.

The following graphs, Figs. 84 and 85, show the graphs of sine, cosine and tangent for the range of values from 0° to 360°.

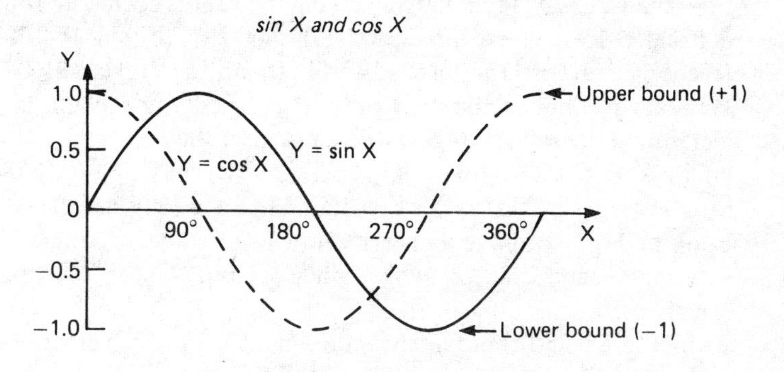

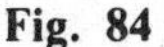
Fig. 84

The cosine curve is of the same form as the sine curve but is 90° out of phase with it.

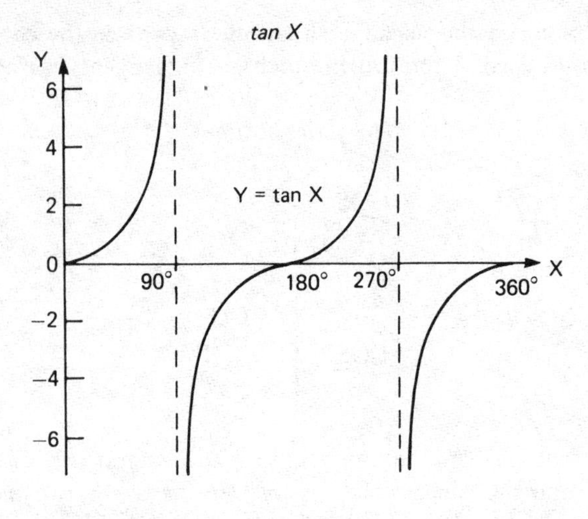

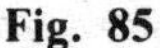

Fig. 85

The vertical dotted lines are the *asymptotes*, the tangents at infinity.

EXERCISE 5

The questions are intended to be for use with an electronic calculator, unless otherwise stated.

1 Find the values of the following giving the results correct to four decimal places, where necessary: (a) sin 123°, (b) sin 164°38′, (c) cos 64°, (d) cos 116°, (e) cos 149°11′, (f) tan 157°, (g) tan 91°07′. What do you notice about (c) and (d)? How do you explain this?

2 Determine the obtuse angles which are such that:
(a) $\tan A = -0.5$, (b) $\sin B = 0.9377$, (c) $\cos C = -0.5878$, (d) $\tan D = -7.9238$, (e) $\cos E = 1.5$ (What is wrong here?).

3 Look at Fig. 84 above and write down the values of angle X, between 0° and 360°, at which $\sin X = \cos X$. Check on your calculator.

4 From Fig. 85, find approximately the angles X for which $\tan X = 2$ and $\tan X = -4$. Compare your results with those obtained by calculator.

5 Find the values of the angles of △ ABC, for which $\tan A = -0.735$ and $\sin B = 0.352$, giving the results in degrees and minutes.

6 *The Cosine Formula.* In $\triangle$ ABC, the sides a, b and c are opposite the angles A, B and C respectively, then $c^2 = a^2 + b^2 - 2\,bc \cos A$. The formula is cyclic, i.e. we can move the letters round the same way ($a^2 = b^2 + c^2 - 2bc \cos A$ and $b^2 = c^2 + a^2 - 2ca \cos B$). The triangle may be acute or obtuse.

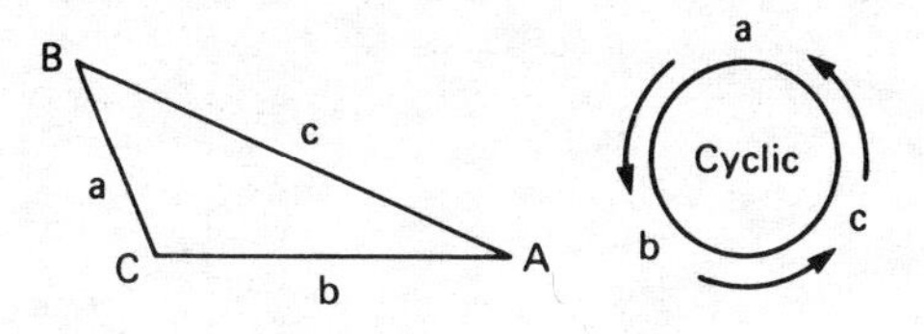

If $a = 27.3$, $b = 38.7$ and $C = 124°$, find c (*note*: $2\,bc \cos A$ means $2 \times b \times c \times \cos A$) correct to one decimal place.

7 *Heron's Formula* for the area ($\triangle$) of a $\triangle$ ABC with sides $a, b\ c$ is

$$\triangle = \sqrt{s(s-a)(s-b)(s-c)},$$

where $s = \frac{1}{2}(a+b+c)$, i.e. the semi-sum of the sides. If $a = 7.4$ cm, $b = 9.3$ cm and $c = 11.7$ cm, find the area of the triangle. (One can obviously work out s by arithmetic and then subtract a, b and c in turn, and then only use the calculator for the last stages, but in fact it can all be done on the machine, as follows:

7	.	4	+	9	.	3	+	1	1	.	7	=	÷	2	=	STO	×

(	RCL	−	7	.	4	)	×	(	RCL	−	9	.	3	)	×

(RCL − 11.7) = $\sqrt{x}$

This key is not really required, as the number is still on display at that moment, but it is a safety precaution as it will certainly be needed for the other brackets.) (Incidentally M + on some calculators takes the place of STO .)

8 *The Sine Formula* for $\triangle$ ABC with sides a, b and c opposite angles A, B and C is

$$\frac{a}{\sin A} = \frac{b}{\sin B} = \frac{c}{\sin C}.$$

If $A = 37° 42'$, $B = 41° 11'$ and $a = 6.8$ cm, find b and c.

Answers

CHAPTER 1

Exercise 1 (page 6)

1 (a) 𒌋𒌋𒁹𒁹𒁹, (b) 𒁹𒌋, (c) 𒁹𒁹𒁹𒁹 𒁹, (d) 𒁹𒁹 𒌋𒌋𒌋𒌋 𒁹𒁹𒁹,

(e) 𒁹 𒁹𒁹𒁹𒁹𒁹𒁹𒁹𒁹 𒁹𒁹

2 (a) 248, (b) 1433, (c) 22 650.
3 (a) 347, (b) 2672, (c) 33, (d) 2020.

4 (a) 𓏺𓏺𓎆𓎆, (b) $3 \times 60 + 21 = 201 \Rightarrow$ 𓏺𓍢𓍢

(c) $2 \times 60^2 + 11 \times 60 + 4 = 7864 \Rightarrow$ 𓏺𓏺𓏺𓏺 𓎆𓎆𓎆𓎆𓎆𓎆 𓍢𓍢𓍢𓍢𓍢𓍢𓍢𓍢 𓆼𓆼𓆼𓆼𓆼𓆼𓆼.

CHAPTER 2

Exercise 1 (page 19)

1 (a) 44, (b) 76, (c) 244. **2** (a) 711, (b) 4181, (c) 63 140.
3 (a) 876, (b) 3877. **4** (a) 132, (b) 355.
5 (a) 2786, (b) 195 605. **6** (a) 91, (b) 384, (c) 7777.
7 (a) 13 r. 49, $13\frac{49}{59}$, 13.83; (b) 20 r. 21, $20\frac{1}{7}$, 20.14;

(c) 79 r. 401, $79\frac{401}{482}$, 79.83.
8 1 1 1 1 1 1 1 1 1; $12345679 \times 8 = 98765432$.
9 (a) 7, (b) 38, (c) 0, (d) 92, (e) 170.
10 529.6 m.p.h. **11** 5 times bigger; 0.45 of the cake.

Exercise 2 (page 24)

1 (a) $3^2 \times 5$, (b) $3^2 \times 5^2$, (c) $2^3 \times 3^2$, (d) $2 \times 3 \times 37$, (e) $3^3 \times 7^2$, (f) $2 \times 3 \times 5 \times 7 \times 17$, (g) $2^3 \times 3^2 \times 13$, (h) $2 \times 3^4 \times 7 \times 31$, (i) 11×23, (j) Does not factorise, (k) 17^2.
2 (a) 12, (b) 14, (c) 33, (d) 3.
3 (a) 45, (b) 72, (c) 2100, (d) 2520.
4 60. **5** 12 noon. **6** 10 min 30 sec; 6 laps.

Exercise 3 (page 27)

1 15 cm. **2** 16 hr. **3** $\frac{3}{4}$. **4** $\frac{43}{47}$. **5** $\frac{181}{146}$.

Exercise 4 (page 31)

1 1740. **2** 400. **3** 27. **4** 0. **5** 19.5 (or $19\frac{1}{2}$).
6 60. **7** 8.5 (or $8\frac{1}{2}$). **8** 96. **9** 6. **10** $\frac{2}{7}$.
11 $\frac{1}{10}$. **12** $\frac{3}{8}$. **13** $-1\frac{13}{140}$. **14** $1\frac{17}{25}$ (or (1.68). **15** 16.
16 $1\frac{20}{69}$. **17** $1\frac{1}{4}$ (or 1.25). **18** $\frac{8}{13}$. **19** £5600.
20 33 ft/sec; 36 km/h.
21 4 days. **22** He must add on $9 \times 79 = 711$ to his result. **23** 72 sec.
24 5; LCM (420; 840; 1260; 1680; 2100). **25** (a) 160, (b) 40.
26 (a) $273 \times 184 = 50\,232$, (b) $19\,293 \div 59 = 327$.

CHAPTER 3

Exercise 1 (page 41)

1 (a) $2xz$, (b) $28y$, (c) $\frac{3}{4}ab$, (d) $\frac{4}{5}abc$.

2 (a) $\frac{2}{3}z$, (b) $\frac{5}{4}y$, (c) $\frac{3ab}{2c}$, (d) $\frac{2}{q}$, (e) $\frac{3b}{7d}$, (f) $\frac{9}{32b^2}$, (g) $\frac{1}{q}$.
3 (a) $x+6$, (b) $11x-6y$, (c) $3a-9b+2c-6$ (terms cannot be collected any further), (d) $2x-10y+12z+5$, (e) -1. **4** $35.2\,n$ oz; $28.4x$ gm.

5 $s-a-b$; $\frac{1}{3}s$. **6** $\frac{3r}{t}$. **7** $\frac{pq}{p+q}$. **8** £$(px+qy)$; £$\frac{px+qy}{x+y}$.

9 £$x-4y$. **10** $\frac{a}{2c}$. **11** (i) $x = \frac{7}{4}$, (ii) $y = 7$, (iii) $z = 14$.

12 $\frac{m}{u}+\frac{n}{v} = \frac{mv+nu}{uv}$ hr; $\frac{(m+n)(uv)}{mv+nu}$ m.p.h.

Exercise 2 (page 48)

2 (a) $x = 5$, (b) $a = 5$, (c) $y = \frac{1}{2}$, (d) $x = 12$. **3** (a) £$\frac{4x+5y}{50}$,

(b) $\frac{8x+10y}{x+y}$ pence. **4** Tea £1.12; sugar 14 p, **5** Father 48; son 12.

6 Breadth $\frac{3c}{2}$ ft; perimeter $16c + 3d$ ft.

CHAPTER 4

Exercise 1 (page 66)

2 $\hat{p} + \hat{q} + \hat{y} = 180°$. **3** $\hat{a} + \hat{b} = 90°$. **4** $\hat{m} = \hat{k} + \hat{l}$.
5 $\hat{d} + \hat{e} + \hat{f} = 360°$. **6** $\hat{a} + \hat{b} + \hat{c} + \hat{d} + \hat{e} = 180°$. **7** $\hat{c} = \hat{a} + \hat{b}$.
8 $\hat{k} + \hat{l} + \hat{m} + \hat{n} = 360°$. **9** $\hat{C} = 93°$; reflex $\hat{C} = 267°$. **10** Hint: use theorem 1.
11 $x = a - c - e$; $y = b + d - f$. **12** (i) $\hat{x} = 32°$, (ii) $\hat{x} = 43°$, (iii) $\hat{x} = 20°$.
13 230° (if to port); 050°; i.e. 320° + 90° − 360° (if to starboard).
14 90°; the figure is a rectangle (we do not have enough information to say whether or not it is a square). **15** 52°, 104°, 156°; 51°.
16 A trapezium. **17** Bearing 100°; distance 63.5 nautical miles (n. mi.).
18 17.3 n. mi at 08.45; 176° at 09.15; not exactly, but they could get uncomfortably close, especially if wind and sea affect their courses.

CHAPTER 5

Exercise 1 (page 77)

1 (a) 3, (b) −11, (c) 11, (d) −1, (e) 0, (f) −10, (g) 4, (h) k, (i) $10y - 3z$, (j) $-3a - 2b + 3c$.
2 $75 + 32 - 47 + 24 - 16 = 68$; 7 points below the initial quotation.
3 (a) 28°C, (b) 7°C, (c) −18°C, (d) −0.6 km.
4 (a) $x = 5$, (b) $y = 3$, (c) $y = 5$, (d) $z = -10$, (e) $y = -\frac{5}{3}$, (f) $z = -4$, (g) $x = \frac{5}{4}$, (h) $y = 7.25$, (i) $x = 2.8$, (j) $x = 6$, (k) $t = 1.3$, (l) $y = -11$, (m) $q = 0$, (n) x is infinite. **5** 5 hr 30 min.

Exercise 2 (page 82)

1 (a) −30, (b) 12, (c) $-\frac{3}{32}$, (d) $-\frac{1}{7}$, (e) $-\frac{1}{3}$, (f) $\frac{3bc}{4d}$, (g) $-\frac{3a}{4c}$.
2 (a) $-\frac{5}{12}$, (b) 72, (c) $-\frac{5}{12}$.

Exercise 3 (page 88)

1 (a) $-4ab - 5a + 3b$, (b) $3abc$ (as $abc = cab = acb$), (c) $xy + 2yz - 2xz$.
2 (a) $2y^5$, (b) $4x^2y^3$, (c) $\frac{15}{n}$, (d) $\frac{a}{4b}$, (e) $\frac{8x^7}{63}$, (f) $14p$, (g) $14st$, (h) $2y^2$, (i) $\frac{1}{3x}$, (j) $\frac{3}{x}$, (k) $\frac{1}{4x^2}$, (l) ± 4, (m) ± 8, (n) $\frac{3}{2}$, (o) $\frac{5}{2}$, (p) $\frac{4d^2}{c^2}$, (q) z^6.

3 (a) $-27, 81$; (b) m^6, m^8; (c) $8a^{\frac{3}{2}}, 16a^2$; (d) $7\sqrt{7}, 49$; (e) $-\dfrac{t^3}{3\sqrt{3}}, \dfrac{t^4}{9}$.

4 $\frac{31}{32}, \frac{63}{64}$; they are nearly equal to 1; if we continue for ever the total is 1.

5 (a) $-\dfrac{x}{12}$, (b) $\frac{19s}{60}$. **6** (a) 13, (b) 20, (c) $\frac{1}{6}$, (d) 9, (e) $\frac{1}{4}$, (f) 3, (g) $\frac{1}{54}$.
7 $\frac{1}{2}$ or -2. **8** (a) -1 or $-\frac{5}{2}$, (b) 1.28 or -0.78.

CHAPTER 6

Exercise 1 (page 96)
1 Congruent (RHS). **2** Congruent (AA cor S). **3** Not congruent.
4 Congruent (AA cor S – *not* RHS). **5** Not congruent: same shape, but not same size. **6** Yes: $\hat{P} = 90° = \hat{Q}$, $PX = b = QY$, $PY = a = QZ$; $\therefore \triangle PXY \equiv \triangle QYZ$ (SAS).
7 In $\triangle$s ABD, ACD, we have AB = AC (given), AD = AD (common line), $\hat{a}_1 = \hat{a}_2$, given (as AD bisects angle BAC), $\therefore \triangle BAD \equiv \triangle CAD$ (SAS); hence $\hat{d}_1 = \hat{d}_2$; but $\hat{d}_1 + \hat{d}_2 = 180°$ (BCD is a straight line) $\therefore \hat{d}_1 = \hat{d}_2 = 90°$.

Exercise 2 (page 107)
2 2.83 in *or* 7.07 cm. **3** SQ = 5.29 cm; angle SQR = 79°. **5** 6 cm.
6 AD = 3.9 cm; angle C = 126°.

CHAPTER 7

Exercise 1 (page 117)
1 21 cm^2. **2** 12.5 in^2. **3** 15; 20 **4** 84; 48 **5** 10.5 cm^2.
6 25 m^2; 23 m. **7** 74 m^2; 34 m. **8** 250 cm^2; 77.02 cm; 32.02 cm.
9 27 × 6 m = 162 m; £129.60. **10** 18.68 m^2; 3.1 litre. **11** 1.84 Ha.

Exercise 2 (page 127)
1 18.375 ft^2; 23.4% (1 dec. pl.). **2** 1.29 m^2; 12.9 m.
4 240 m^2; 10.91 m. **5** 1350 cm^3; 1008 cm^2. **7** In q. 5, $F_n + V_n - E_n = 5 + 6 - 9 = 2$ (yes); in q. 6, $F_n + V_n - E_n = 4 + 4 - 6 = 2$ (yes).

CHAPTER 8

Exercise 1 (page 143)
1 420. **2** 2271. **3** 139. **4** 94. **5** -7.
6 608.339; 608.34. **7** 0. **8** 29 184. **9** 98.4903; 98.49.
10 1541.0304; 1541.03. **11** 20.28. **12** 22.506 486; 22.51.
13 0.001 574 88; 0.00. **14** -1.6. **15** 0.8.
16 0.834 920 64; 0.83. **17** 0.835 779 18; 0.84.
18 6.899 275 3; 6.90. **19** 10.5 (in.). **20** 0.010 187 32; 0.01.
21 1.791 324 8; 1.79. **22** 7099. **23** 12.28; 12.280.

24 $-0.734\,984\,52$; -0.735. **25** 1.828 282 8; 1.828.
26 60.082 707; 60.083. **27** 0.344 696 97; 0.345.
28 0.17; 0.170. **29** $-0.143\,120\,96$; -0.143.
30 91.356 392; 91.356. **31** 4.004 085 4; 4.004.
32 0.543 855 3; 0.544.

Exercise 2 (page 148)

1 (a)

1	9	×	7	C	6	=

114 (illustration); (b) 23.

2 32. **3** -10.613.
4 $8.7106 \times 10^9 \simeq 8.711 \times 10^9$. **5** 6.8216×10^9.
6 $31\,536\,000 = 3.1536 \times 10^7$; 1.2623×10^8. **7** 2.79×10^{-5}.
8 1.282 sec (3 dec. pl.); 5.37×10^{-6}. **9** (i) 625, (ii) 7776, (iii) 392.866 51, (iv) $2\,985\,984 \simeq 2.986 \times 10^6$, (v) $2.565\,784\,5 \simeq 2.566$.
10 (i) 2.303 230 6, (ii) 0.760 551 11, (iii) 233.908 25.
11 (i) 0.1978, (ii) 0.1317, (iii) 1.7014; this is a little tricky and is therefore illustrated

1	1	÷	7	=	y^x	3	=	EXC	8	÷	9	=	y^x	7	=	×	RCL	=

(iv) Easiest way

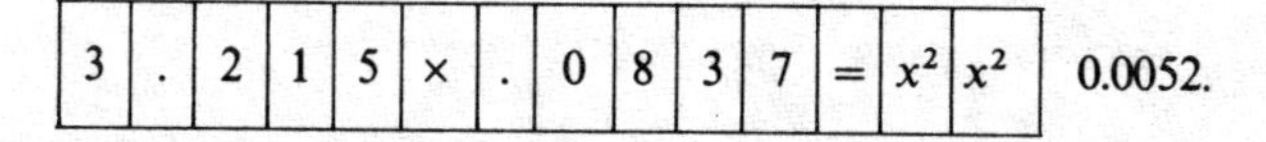

12 (i) 2.074×10^8, (ii) 2.056×10^4. **13** (i) 3.9×10^{-7}, (ii) 3.855×10^{-13},(iii) 5.751×10^{-10}, (iv) 6.43×10^{-5}.
14 0.00054113 ($= 5.4113 \times 10^{-4}$).

Exercise 3 (page 153)
1 -294. **2** 39.6836. **3** 9.608 571 4.
4 $-2.741\,946\,4$. **5** -1856 (to nearest integer). **6** -0.937.

CHAPTER 9

Exercise 1 (page 160)
1 28.27 cm. **2** 19.5 cm; 298.65 cm^2. **3** 35.45 cm.
4 113.1 cm^2. **5** 856 cm^3. **6** 4.775 cm; 1647 cm^3 (nearest 1 cm^3).
7 55.2 cm^3; 67.2 cm^2. **8** (i) 12 cm, (ii) 314 cm^3, (iii) 283 cm^2.
9 (i) 32 cm, (ii) 31 cm, (iii) 3124 cm^3.
10 (i) 215 ft^2, (ii) 245 ft^3, (iii) 4 hr 14 min (to nearest minute).

11 $V_2 : V_1 = (\frac{1}{3})^3 : 1 \therefore V_2 = \frac{V_1}{27} = \frac{20}{27}$ cm³; calculator not needed, except for decimals.
12 940.2 cm². **13** 58.7 cm³. **14** 697.1 cm²; 1332.2 cm³; 10.5 kg.
15 20.7 m³; 43.4 m². **16** 447 600. **17** 37.06 cm³.
18 3.7385 m³; 27.7 tonne. **19** (i) 22 734 m³, (ii) $4a\sqrt{b^2 - a^2}$, (iii) 3655 m². **20** 11.06 cm. **21** $\frac{5}{6}r, \frac{25}{36}\pi r^2; \frac{5}{6}\pi r^2 = 167.6$ cm².

CHAPTER 10

Exercise 1 (page 167)
1 (i) 0.25, (ii) 0.364, (iii) 0.075, (iv) 2.53.
2 (i) 5.8824 %, (ii) 12.5 %, (iii) 83.6 %, (iv) 502 %, (v) 81.6497 %.

3 0.7956 %

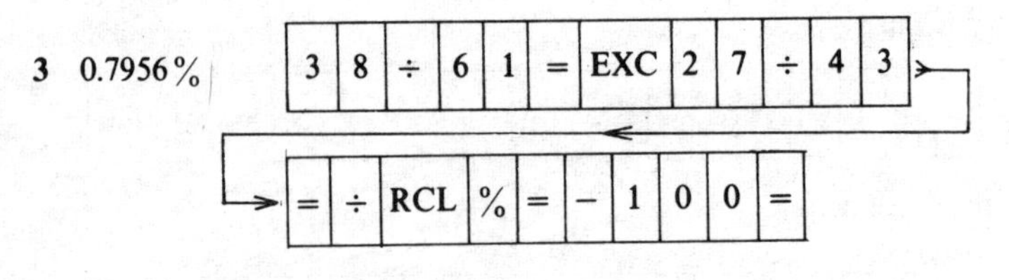

4 (i) £362.67, (ii) 620.8 cl, (iii) 26p. **5** 6.25 %

Exercise 2 (page 169)
1 £28.14. **2** £9.89. **3** (i) 34.43 %, (ii) −2.51 % (2 dec. pl.).

4 £93.5. **5** 40 %. **6** £96.98

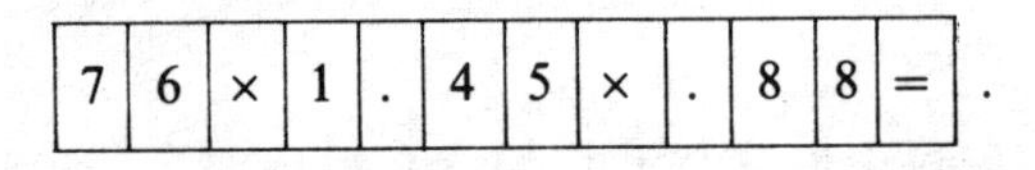

7 95p (nearest 1p). **8** £16.68. **9** (i) £275.20, (ii) 21.5 %, (iii) No.

Exercise 3 (page 172)
1 (i) £1146.60, (ii) £1210.32. **2** 5 years. **3** £3048. **4** $12\frac{1}{2}$ % (calculator not needed). **5** £95.12.

Exercise 4 (page 176)
1 £842.70; £92.7. **2** £4328.41; £868.41. **3** £7108.11; £2204.43.
4 £26 172.95; £8329.22. **5** £334.71. **6** £357.52; solution

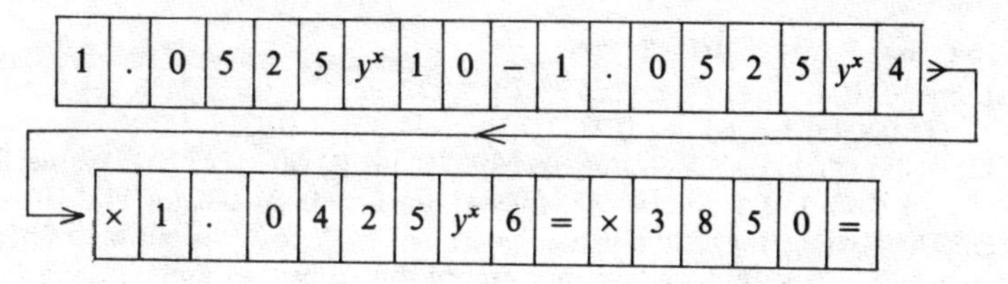

for most modern calculators having y^x key.

7 £250; $\frac{250}{1897.48} \times 70\% = 9.22\%$.

Exercise 5 (page 182)

1 E = 220, $P_5 = 2000 - \frac{100 \times 220}{14}\{(1.14)^5 - 1\}$ pounds = £545.78 (nearest 1p).

2 (a) E = 180 − 180 = 0 ∴ mortgage remains at £18 000;
(b) E = 240 − 180 = 60, $r = \frac{12}{12}\% = 1\%$, $n = 60$ (monthly instalments), hence $P_{60} = £18\,000 - \frac{100 \times 60}{1}\{(1.01)^{60} - 1\} = £13\,099.82$;
(c) E = 300 − 180 = 120, hence actual monthly repayment, *after* deducting interest, is doubled ∴ saving over 5 years is double what it was in (b) above. Hence, outstanding mortgage is
£{18 000 − 2(18 000 − 13 099.82)} = £8199.64,
thus avoiding another application of the formula above.

3 Up to 53 months: $P_0 = 16800$, $E_1 = 36$, $r_1 = 1.125\%$ (monthly) ∴ $P_{53} = £14\,210.35$. For remaining time: $P'_0 = 14\,210.35$, $E_2 = 100.659\,44$, $r_2 = 0.875\%$, $n_2 = 96 - 53 = 43$. Leading to P_{96} = £8982.63 (outstanding mortgage at the end of 8 years from starting).

4 £9.97.

CHAPTER 11

Exercise 1 (page 188)

1 DE = 4 cm, DF = 5.6 cm. **2** (ii) DE = 5 cm. **3** 3.4 cm and 4.3 cm.

Exercise 2 (page 195)

1 (a) 0.0875, (b) 0.5543, (c) 1.0356, (d) 4.7046, (e) 0.2586, (f) 0.6519, (g) 1.1067, (h) 2.6348, (i) 2.7044, (j) 1.5311, (k) 0.4, (l) 2, (m) 0.0015, (n) 114.59.

2 (a) 35°, (b) 81°, (c) 15° 30′, (d) 59° 60′, (e) 44° 54′, (f) 8° 16′, (g) 34° 16′, (h) 49°, (i) 77°36′, (j) 89° 20′, (k) 45°. **3** (a) 27.00, (b) 6.40, (c) 3.26, (d) 45.63, (e) 0.00, (f) 172.87, (g) 94.42, **4** · 83° 33′.

5 (a) 16.2, (b) 6.1, (c) 4.8. **6** 6.4 m. **7** 56° 27′. **8** 1 m 21 cm.

Exercise 3 (page 202)

1 (a) 0.7071, (b) 0.7071; $\sin 45° = 1 \div \sqrt{2}$, (c) 0.5, (d) 0.0436, (e) 0.9965, (f) 0.8, (g) 0.4365, (h) 0.9878, (i) 0.5974, (j) -0.9527.

2 (i) (a) 0°, (b) 14° 29′, (c) 36° 09′, (d) 75° 14′, (e) 90°, (f) no real value, for $-1 \leqslant \sin A \leqslant 1$; (ii) (a) 90°, (b) 84°, (c) 0°, (d) 50° 08′, (e) 118° 10′.

3 The obvious suggestion is that $\sin 2Y = 2 \sin Y \cos Y$ for all values of Y, which is true (but we have not *proved* this to be the case).

4 Each is equal to $-0.327\,2179$; we conclude that $\cos 2A$ *may* equal $\cos^2 A - \sin^2 A$ for all values of A (which is in fact true).

Exercise 4 (page 208)

1 (a) 16.22, (b) 4.31, (c) 3.37, (d) 43.59, (e) 0.87, (f) 4.94, (g) 1.17, (h) 2.

2 (i) $x = 10.23$, $y = 10.97$; (ii) $\cos Y = \frac{8}{14}$ $\therefore Y = 55° 09'$, $y = 8 \tan Y = 11.49$; (iii) $x = 6 \sin 32° = 3.18$, $y = 6 \sin 32° \tan 41° = 2.76$.

3 $BC = 33 \tan 35° \simeq 39.40$, angle CAD $\simeq 4° 24'$.

4 $\cos Z = 5 \div 12.5$ $\therefore$ $Z = 66° 25'$; $YW = 5 \cos Z = 2.00$.

5 Using Pythagoras' theorem, $AM = \sqrt{AB^2 - BM^2} = \sqrt{4-1} = \sqrt{3} = 1.7321$; $\sin 30 = \frac{1}{2} = 0.5$, $\cos 30° = \dfrac{\sqrt{3}}{2} = 0.8660$; $\sin 60° = 0.8660$, $\cos 60° = 0.5$.

6 The height is $25 \div (\tan 58° 16' - \tan 51° 43') + 1.5 = 71.4 + 1.5 = 72.9$ m.

Exercise 5 (page 216)

1 (a) 0.8387, (b) 0.2650, (c) 0.4384, (d) -0.4384, (e) -0.8588, (f) -0.4245, (g) -51.30; (c) and (d) have the same value because $64° = 180° - 116°$.

2 (a) 153° 26′, (b) $180° - 69° 40' = 110° 20'$, (c) 126°, (d) 97° 12′, (e) cosine lies in the range -1 to $+1$; 1.5 is outside this range for real values of this function.

3 45° and 225°. **4** When $\tan X = 2$, $X = 63°$ ($63.43° \simeq 63° 26'$) or $X = 63° + 180° = 243°$ (243° 26′); when $\tan X = -4$, $X = 1.15 \times 90° = 103.5°$ (104° 02′).

5 This requires a little care: $\tan A = -0.735$ can only give angle $A = 143° 41'$ in a triangle but this is obtuse, hence B and C must both be acute, because all 3 angles add up to 180°; $\therefore$ as $\sin B = 0.352$, then $B = 20° 37'$ (and *not* 159° 23′) $\therefore$ $C = 180° - A - B = 15° 42'$.

6 58.51 (the cyclic form needed was, of course, $c^2 = a^2 + b^2 - 2ab \cos C$).

7 34.39 cm². **8** $b = 7.32$ cm, $c = 10.91$ cm.

ALGEBRA

P. ABBOTT
Revised by MICHAEL WARDLE

This straightforward introduction outlines the principles and foundations of algebra with particular emphasis on its applications to engineering and allied sciences.

Covering the ground from the very beginning, the text explains the various elements of algebra such as equations, factors and indices, continuing on to simple progressions and permutations. The course is carefully graded and only a knowledge of elementary arithmetic is assumed – included in each chapter are a number of exercises (with answers) designed to test and encourage the reader's progress.

TEACH YOURSELF BOOKS

CALCULUS

P. ABBOTT
Revised by MICHAEL WARDLE

This well-known introduction to calculus, now fully updated to include numerical work, enables the reader to tackle a range of problems within pure and applied mathematics, engineering and allied sciences.

Ideal for beginners, the book provides a carefully-graded series of lessons which introduce the basic concepts of differentiation and integration. Each chapter includes many clearly-worked examples, diagrams and exercises with answers.

Michael Wardle has added new material illustrating numerical approaches to the subject using a calculator and showing a range of applications, including the numerical solution of differential equations, which are possible on a home computer.

TEACH YOURSELF BOOKS